UNFINISHED NATURE

UNFINISHED NATURE

PARTICLE PHYSICS AT CERN

ARPITA ROY

Columbia University Press   *New York*

Columbia University Press
Publishers Since 1893
New York   Chichester, West Sussex
cup.columbia.edu

Library of Congress Cataloging-in-Publication Data
Names: Roy, Arpita (Anthropologist), author.
Title: Unfinished nature : particle physics at CERN / Arpita Roy.
Description: New York : Columbia University Press, [2024] |
Includes bibliographical references and index. Identifiers: LCCN
2023033784 | ISBN 9780231205528 (hardback) | ISBN 9780231205535
(trade paperback) | ISBN 9780231556040 (ebook)
Subjects: LCSH: European Organization for Nuclear Research. |
European Organization for Nuclear Research—Employees—Social
conditions. | Nuclear energy—Research—Laboratories—
Sociological aspects. | Nuclear physics—Research—Social
aspects—Switzerland. | Discoveries in science—Social
aspects—Switzerland. | Higgs bosons.
Classification: LCC QC789.2.S9 R69 2024 | DDC 539.7/3—dc23/
eng/20231122
LC record available at https://lccn.loc.gov/2023033784

Cover design: Milenda Nan Ok Lee
Cover image: vchal/Shutterstock

Inclinata resurgit

CONTENTS

ILLUSTRATIONS

FOREWORD

Four Riddles of Physics and Epistemology

MICHAEL M. J. FISCHER

Arpita Roy provides the latest of a series of ethnographic ventures into high-energy physics, in this case
primarily at CERN. Coming in part from mentor-
ship of J. P. S. Uberoi, the eminent anthropologist-philosopher
at the Delhi School of Economics, she attempts to revisit a
number of old and new riddles that physics poses for general
epistemology, trying to get beyond the presuppositions with
which we conventionally talk about physics and nature. Earlier
variants of this desire have been provided by the juxtapositions
of cultures with other presuppositions or with other traditions
of scientific discovery, whether in anthropology or in the history of science (as in Uberoi's rereading of Goethe's science as a
counter tradition in Western and global science). Roy attempts
to do this more rigorously through living with and debating
physicists as they try to explain their work.

Chapters 3 through 5 provide fascinating case studies. Chapter 3 is about left- and right-handedness (of the body, of the spin
of particles), an old anthropological topic since the studies of
Robert Hertz in the overlapping terrains of what are nature, culture, natural symbols, symmetry, the limits of topological transformations, and perhaps the challenges of viewing biochemical

cascades in multidimensional space. Chapter 4 is about the 2008 explosion in the Large Hadron Collider, which, in its forensics and reconstructions, provides exposure of the materialities of highly speculative science experiments. It is an old verity in engineering that one often learns as much from the failures of systems as from when they work according to plan; in any case, failures help demystify the shiny surfaces and secrets of complexity, revealing the nature of work and imagination. Chapter 5 is about CERN's efforts to allow access to artists-in-residence. This is no longer as novel an experiment as it seemed some thirty years ago, and there are now many art-science projects, which, when most productive, are not just metaphorical publicity to support the arcane mysteries of technical knowledge for the general public but are also experimental spaces where some of the constraints of science can be relaxed to see what happens and if, thereby, new insights can spark the scientific imagination.

The book begins in chapters 1 and 2 with the excitement surrounding the discovery of the Higgs boson in 2012, a long-anticipated international quest of theoretical collaborations in international physics, and the building of such transnational experimental machines as the Large Hadron Collider, which are beyond the means of national governments but are none-theless matters of competition to provide centers of physics activities. CERN was part of the post–World War II effort to rebuild Europe's scientific infrastructures and political establishments. The Higgs boson, like many of physics' high energy particles, is a weird object: although evidence of Higgs bosons have been "found" or "produced," Roy tells us, they can never be directly observed in particle detectors that produce the evidence for them, partly because they decay so rapidly into two photons and partly because they are hard to differentiate from background photons.

Readers will bring to chapters 1 and 2 different backgrounds and their own understandings of the philosophy of science, which may make Roy's arguments more or less difficult to accept, and I suggest that after a brief review (which I will try to present below), one delve into chapters 3 through 5 before attempting to come back and clarify, in one's own terms, chapters 1 and 2. Chapters 3 through 5 can just be read on their own. But they too gain something from the most interesting challenge Roy poses in chapters 1 and 2 stemming from her observations about the finding/production of the Higgs Boson.

We recognize the Higgs Boson by its "signature," "models anticipating its existence," human judgment (of what are bosons and what background), and significance. It is this last, in particular, that is most interesting (and least predictable). What role, Roy asks, "will the signature or discovery of the Higgs particle play in the next few decades?" At issue, in conventional terms, are the Standard Model, or new frontiers in the work of physicists, terms that are of primary interest to physicists. But for the rest of us, perhaps the interest is rather the ways in which science is always open-ended. *Unfinished nature* is Roy's term for the open-endedness of scientific discovery. It resonates with my own "Emergent Forms of (Un)Natural Life"—explorations in synthetic and systems biology that provide new arenas of similar/different epistemological issues as particle physics. Similar/different is what those physicists learned who entered biology in the mid-twentieth century in the hopes of "solving" the mysteries of the genetic code as if they were cryptographies. As the historian Lily Kay demonstrated, the "code" proved resistant to them and was "cracked" instead by microbiologists, illustrating, as Kay put it, that genetics was not simply a code but complex cascades of regulatory systems that often mutate and generate the new: genetics and genomics had to be expanded into many different "omics" if we were going to understand much about biology (and indeed ecology, habitats, and human and nonhuman interactions). Hans-Jörg Rheinberger would call the experimental method in biology a generator of surprises, more like the open systems of Ludwig von Bertanlanffy and the deconstructive reading and writing of Jacques Derrida, whose work Rheinberger both translated and used in his account of the discovery of protein cascades.

RIDDLE 1: SIGNATURES AND WHAT ROLE WILL THE HIGGS BOSON PLAY IN THE NEXT FEW DECADES?

Roy writes, "On July 4, 2012 . . . two of the [four] collaborations at CERN, ATLAS and CMS, announced that they had discovered a new particle in the mass range of 125 gigaelectron volts (GeV) consistent with the Higgs boson predicted by the Standard Model." This was popularly glossed as the discovery of the Higgs particle. Riddle 1: as Roy nicely puts it, while anticipated, "it is not foretold what precise role it will play in the next few decades" even as the researchers who discovered it won the Nobel Prize in 2013. Indeed, what is its significance?

The Higgs boson, an "elementary" particle theorized long before it was experimentally found, was one of the great international quests at the Large Hadron Collider (LHC) at the CERN particle accelerator, built in 2012 to consolidate European scientific infrastructures in quasi competition and quasi cooperation with accelerators in Japan and the United States. It was, like the earlier accelerators at CERN *a political and political economy project* as well as a scientific one—*and an epistemological one.* One can count "generations" of precursor scientific experimental apparatuses that involved both different ideas about how to "tinker with" versus "machine-tool" high-precision instruments (Traweek 1988) and different epistemologies about how particles can be identified (Galison 1997). Traweek contrasts American physicists, who are often farm boys used to repairing and inventing machinery, who believe int their personal tinkering abilities; and Japanese physicists from more upper class backgrounds who depended upon machinists to work for them. Galison provides a history of the competition between physicists who privileged a singular visual trace (a "golden event" or image) and those who privileged statistical pattern recognition (or "logic").

For some 'seeing is believing' continued, but a golden event can't show you anything (at best is an indexical trace); statistics can tell you things more robustly, but you can't see them.

Even this initial bit of twentieth-century history of high-energy physics has given sociologists and anthropologists material to think about how science is patterned in different historical, cultural, and epistemic traditions. Two key moments in this initial history are Sharon Traweek's closely detailed account of physicists in two training traditions that simultaneously seems to reflect larger industrial patterns, with Japanese physicists relying on machinists, just as Japanese small supplier firms power their larger international industrial ones (ethnographically described by Kondo, 1990); and Galison's account of the development of instruments in particle physics from "bubble chambers" modeled on alpine cloud chambers or visible models of dynamics of patterns, not unlike wind tunnels or wave machines, evolving into tracks left on "detectors" of "invisible" events whose logic (and reality) is demonstrated by their statistical "signatures." Galison shows how, slowly, with competition between the epistemologies and capacities of different instruments, what eventually clinched belief in truth (of a model) or existence (of a particle) merged:

logic and image coalesced; theory and experiment overlap and interrogate each other. Knorr-Cetina adds a useful contrast within this ethnographic tradition, arguing (along with Galison) that physics has become inextricably the work of large-scale teamwork (and historical sociologists like Mario Biagioli would show how the politics of citation worked in apportioning credit among team members for, among other things, professional promotion, fund raising, and access to machines), in sharp contrast to the continuing role of the scientists' own "good hands," intuition, and "body" in molecular biology, a field that since then has also rapidly become statistical big science. .

Arpita Roy brings us back to some basic questions about signatures, models, and realities. Although Higgs bosons have been "found," Roy tells us, they can never be directly observed in particle detectors, in part because they decay so rapidly into two photons and in part because they are hard to differentiate from background photons. The background photons are given off as electromagnetic radiation by accelerated particles when forced to orbit in the circular trajectory of particle accelerators, used to generate bosons in high-speed particle collisions. This differentiation between the photons in boson decay and the background photons, she says, is a *matter of judgment* by physicists. The experimental detection of the Higgs boson can be done only indirectly, through other particles produced in wake of its decay such as photons, leptons, and hadronic jets. So, are the bosons objectively real? Here the language around the issue becomes a bit philosophically fuzzy. As she quotes Ian Hacking, the question is ill-formed: as any pragmatist, logical positivist, or operationalist (to name just a few of the overlapping reflections of physicists about the philosophy of what they are doing) would say. *Of course it is real, as long as it does productive work.* (To call this "socially constructed," as sociologists of science of a certain stripe, faddish in the 1990s did, is not a very helpful substitution of jargon, providing a false or merely linguistic sense of precision.) Euclidian geometry was socially constructed and was considered universally true, until the presuppositions were changed and non-Euclidean geometries became not only theoretical potentials but also practical tools. It is this *level of presuppositions* that Roy wishes to access, and an account that does so would be quite valuable. Unfortunately, the complications of using ordinary English to describe her concerns lead into old-fashioned formulations such as intrinsic versus extrinsic relations in science (e.g., logic or verification through experimentation versus funding and political patronage). Roy describes her ambition "to disclose not so much the discovery of a new fact, as the discovery of new ways of thinking about a fact." She attacks an old-fashioned (and mischaracterized) variant of ethnography as naïve description as if it was finished fact-finding and insists on the inescapability of any ethnographer today that "fieldwork among a community means attending to persons and things, learning from them, and re-uniting knowledge." This is not "beyond the pale of description," as she claims in the introduction, but a necessary component of it. (And has been so since at least the 1920s, if not back to Vico and Fustel de Coulange.) The credibility of Roy's comments about physicists' understanding of their own presuppositions depends upon our being able to credit her accounts of ethnographic method as well. She writes, "One of the great puzzles for sociologists and philosophers is to explain how scientific discoveries and innovations constitute a break from a given empirical context." This is the old issue of how models function between logic and phenomenology. In anthropology, perhaps most productive were the distinctions between etic and emic perspectives (provided by linguistic analysis) or between generative deep structures and ordinary surface structures as in the classic example of language speakers who recognize correct versus incorrect grammar but cannot provide the rules. In physics the riddles (Roy's nice term) have more to do with the logics of mathematics (of various kinds and presuppositions) and the conventional realities or agreements among practitioners.

She writes, "A fundamental premise of this book is that the context of discovery in particle physics cannot be explained by any compilation of empirical results since it expresses itself in the heuristic of conceptual possibilities," involving always the relationship between the framing of problems and the "facts" gathered to support or disconfirm them. This is true not only in physics. The production of a topology of conceptual possibilities might be an enlightening project, but if that is what Roy has in mind, she doesn't quite get there.

She nicely quotes Marx on the Greek arts and epics as bound to social forms of development (their time and place), yet they "and still give us (who are located in other times and places) not only artistic pleasure, but norms and models that cannot be rivalled." Roy sees this as disjunctive; I do not. Indeed, as she says in the following sentence, mistaking this "riddling problem" as distinctive of science (I would think it distinctive of aesthetics), how natural science (or aesthetics) "is integrally bound up with the social milieu, while it preserves its distinctiveness and promotes the expansion of knowledge, which furthers discovery and invention and even experimentation."

Roy wishes to argue that scientists are (usually) unaware of the presuppositions that they operate with, but Roy operates with her own presuppositions, especially that one can code scientific presuppositions as having relations of fact/value, subject/object, and theory/practice. That these words are everyday usages seems empirically true, but suppose that the term "subject" includes the engineering apparatus and one comes close to the Heisenberg indeterminacy principle that her (Roy's) presuppositions refuse. This is a riddling problem (if not a mistake) on the part of the ethnography (not the science) or of the descriptive account, and perhaps of the theory of binaries—where do binaries come from, if not from always changing linguistic discriminations? Indeed, even she says that "the power to make things real and active belongs to the domain of concepts," i.e., linguistic tokens, which go in and out of fashion. It may certainly be true that, as Thomas Kuhn tried to argue, concepts are smaller units locked into paradigms of thought and paradigms change only when better ones (able to explain more facts with fewer anomalies) are devised (note the active human verb "devise," as in devising a play out of actors' multiple experiences and interpretations or devising an experiment out of multiple scientists' earlier experimental schemes). Of course, in science, purely subjective feelings, perceptions, and sensations are marginalized, *but not as a matter of discovery*, only as matters of verification by multiple scientists. At times she wants it both ways: "There is no reason why we should be coy in approaching the technical concepts to discern features of subjectivity or contingency, and then we may be fully justified in regarding natural science as a normative activity or social construct like any other."

Roy wishes, as this last sentence indicates, to bridge the "gap" between C. S. Lewis's two cultures (the sciences and the humanities), an old topic of discussion from the late nineteenth century, perhaps best overcome in Wilhelm Dilthey's notion of intersubjective and dialogic communication, making all thinking public, not private or hidden, open therefore to modes of confirmation or reframing (socially constructed, if you insist, while growing from the "internalist" developments of previous work and also using various kind of pidgin languages where different fields of expertise with their different conventions must find ways of communicating).

In this context, then, "riddling problems" is not a bad frame for challenging scientific problems, both as they play out conceptually and as they are channeled by their funding, accrediting, and other (sometimes frozen and sometimes reassorting) social environments.

RIDDLE 2. LEFT HAND, CHIRALITY, CHIROPTICAL OBJECTS, AND SURREALISM

Chapter 3 might be read as a symphony integrating four different sorts of melody or logic: left-right symbolism; chirality (a geometrical property that an object or structure and its mirror object, or enantiomer, are not superimposable by any translation or rotation, something Leibniz tried to address with his topology); chiroptical phenomena (rotational features that are quantifiably measurable unlike the pure geometrical chirality); and René Magritte's (or other surrealists') provision of an epistemological parallel in visualization methods.

It is a stimulating tour through the boundary land between commonsense puzzles of left-handed gloves and other such asymmetries (while right and left hand look the same, a left glove cannot fit on a right hand, except if it is turned inside out), on the one hand, and, on the other, more fundamental findings by physicists that elementary particles have spin (or are left- or right-handed or spin clockwise or counterclockwise, also called negative and positive [right-handed] helicity). In particle physics this asymmetry is further made puzzling by the fact that while for every particle there should be an anti-particle ("symmetry"), in fact after the big bang, more left-handed particles were produced. (One theory is that as the universe expanded it cooled and couldn't keep up the symmetry of particle–anti-particle pairs, leaving an asymmetry of more left-handed ones carried by weak electromagnetic forces). Richard Feynman and others have suggested this might be a future way of communicating to aliens elementary notions of orientation, so important in biology on earth.

Chirality (handedness) is fundamental to cells, DNA, and proteins and is an important feature in artificial nanoparticles and new materials. What is so interesting about this asymmetry (from the micro level of particles to the relatively macro level of asymmetry in biology such as right and left hands) is the geometrical property that an object or structure and its mirror object (enantiomer) are not superimposable by any translation or rotation. Leibniz tried to construct a topology to deal with this issue (identified by Kant, that the orientation of the two hands could not be reduced to metrical properties). Chirality is a geometrical feature, but biological processes also are constituted by *chiroptical* properties, measurable by chiroptical spectroscopy in various kinds of rotation, dispersal, refractions, polarization, and phases. These help constitute embryological processes defining organ orientation and placement.

Roy, appropriately given her fieldwork at CERN, pays most attention to the puzzles of chirality and provides a short history of the experiments of the early post–World War II period "atom smashers" in laboratories at Berkeley (the bevatron), Brookhaven (the cosmotron), Birmingham (the synchrotron), and CERN (the proton-synchrotron). These machines produced subatomic particles with which theories about the beginnings of the universe could be explored, a program that the search for the Higgs boson continued. But she also intriguingly hints at a Lévi-Straussian (or structuralist) account of the near universality of left-right symbolism in cultural systems, what anthropologists once suggested might be "natural symbols." A too-easy solution to natural symbols would be to say that physicists have provided a natural substrate to explain the asymmetries (what is sinister or left and what is right or moral); a more likely solution, as Roy suggests, is the idea of holistic systems of symbolic tokens, part of linguistic system formations themselves.

In a very nice final move in the chapter, Roy allows the physicist Luis Álvarez-Gaumé to guide her in a discussion of René Magritte's 1937 painting *La reproduction interdite*, meaning ambiguously either "reproduction forbidden" or "impossible reproduction," the latter being more apt. It is a painting of a man standing in front of a mirror, but instead of seeing his front in the mirror, we see a second slightly awry variant of his back. To his right on a table is a copy of Edgar Allan Poe's novel *The Narrative of Arthur Gordon Pym of Nantucket*. The book is painted realistically, as we would expect to see it, unlike the man and his image, but the novel is full of the quirks of an unreliable narrator. It is as if Álvarez-Gaumé is acknowledging the weirdness of particle physics, with its breaking of normal symmetries, parities, and the like, and indeed with remaining open questions, dependent on things we cannot see directly (requiring indirect observation, cleaning of data, and refining of models and theories). It is a bit odd that in chapter 5 Álvarez-Gaumé's openness to art-science synergies of different but mutual exploration becomes attenuated if not denied: it is as if he has split selves or has stepped into Magritte's painting.

RIDDLE 3. NUTS AND BOLTS: TUNNEL EXPLOSION AND REPAIR

On September 19, 2008, just nine days after the new LHC had been inaugurated, a gas explosion suspended operations for many months, amid intense competition with the Tevatron at Fermi National Accelerator, in Illinois, and fears that this set-back might cost CERN the chance to find/produce the evidence of the Higgs boson first. The explosion released two tons of helium gas, breaking magnets and displacing others from their moorings. The liquid helium containers, which were punctured by an electrical arc when an interconnect splice between mag-nets disintegrated, were supposed to keep the magnets and tun-nels at a 1.9 Kelvins (or -271° Celsius) operating temperature. After the accident, it would take time to bring the tunnels up to room temperature to allow for repairs.

Chapter 4 is, or can be read as, a real "case study" of industrial failure, or as Charles Perrow classically formulated it, "natural accidents," involving more or less "tightly coupled" social com-ponents of complex engineering projects. The STS scholars Diane Vaughn and Constance Perrin followed up regarding the 1986 Space Shuttle *Challenger* disaster and the 1979 partial melt-down of the Three Mile Island nuclear plant and other brittle nuclear safety cultures. Vaughn focused attention on the internal power relations among engineers, administrators, and manu-facturers that allowed for a managerial risk decision to be taken that proved fatal. The physicist Richard Feynman became the iconic figure in the forensic inquiries and hearings by simply holding up a rubber "O" ring that failed (it could not expand in 32 degree weather as the rocket heated up) and should have been recognized as likely to fail (as engineers warned), and that ipso facto threw harsh light on the rosy probabilities of failure and safety that NASA used in its public relations. Constance Perrin attempted to provide nuclear power stations with "safety cultures" that have been pioneered in other critical mission are-nas such as surgical anesthesia and airplane cockpit design. She ran a two-year seminar in the MIT STS Program to explore these safety cultures and their challenges while doing fieldwork in nuclear facilities in the United States and Europe. Among her warnings, again, was a sociological one, that a primary challenge needed to be to increase the relative prestige of safety officers in the plants, whose status was normally at the bottom and whose recommendations were treated as routine and unimportant housekeeping. The industrial accident at Bhopal in 1984 and the failure at Chernobyl in 1986 have become classic cases in what the sociologist Ulrich Beck called "risk society," a second-order systemic formation in modernity that works to remove the "means of perception" from individuals and hides risks among a variety of bureaucratic processes, not unlike the way in which Franz Kafka analyzed how mining industries hid their risky practices from the view of insurance companies so that they could shield accidents and death from liability. The failure of the Daiiche Fukushima nuclear plants in the 2011 tsunami was largely caused by misrecognizing that heights of tsunamis in the future would increase and placing the backup generators in basement rooms rather than higher in the facility.

Accidents such as these and others become case studies for the fragility of complex systems but also moments exposing the competition and relative standing and power among theorists, experimentalists, and instrument makers. This is a classic framing often attributed to Thomas Kuhn (does theory lead science, or does experiment?), then refined in the history of twentieth-century physics by Galison's inserting the importance of instruments (as described above). But accidents also force the recognition that more players are involved and could be critical positive safety monitors, including regulators, policy managers, and even ethnographers. It is indeed in the postaccident debates, grumblings, and corridor complaints that Roy says she first truly understood the sharp divisions between getting back to work (theory), getting the instruments up and running, and rethinking experimental strategies (especially timing of repairs and experiments to not allow the Americans to get ahead). The analogue to the *Challenger*'s O ring here was the disintegration of an interconnect splice in turn attributed to the mundane (if crucial) lack of care in soldering. Among the riddles Roy's work presents in this chapter is what effect a Diane Vaughn, Constance Perrin, or Charles Perrow might have had if their forms of attention and accountability checking been in place? As we learn from such case studies, these are not simply technical matters (best practices, check lists, maintenance checks) but also matters of power competition among many different kinds of actors, who on the surface all seem to want to work in seamless cooperation toward common ends but whose assertion of interests throws obscuring shadows over equally important but less powerful ones. (No science and engineering without the peopling of these fields.)

RIDDLE 4. NULL RESULTS: EXPERIMENTAL AND ETHNOGRAPHIC FAILURES, REVISING THEORY

In 2000 CERN created several three-month residencies for artists. Such residencies have become quite popular in various science and engineering settings over the past several decades. The expectations and outcomes can be quite varied, and one of the affordances is that expectations are kept loose. Still, as with laboratory experiments, experiments in the arts do not always live up to expectations. Roy's one-time advisor at Berkeley, Paul Rabinow, once wrote a controversial book on his ethnographic failures in Morocco in which he questioned the entire enterprise as a colonial misadventure. He subsequently redeemed himself, first, by expanding his field of vision to the nature of French colonial modernity and, later, by a finely crafted ethnography of a molecular biology laboratory and the making of PCR (polymerase chain reaction), which was becoming a fundamental laboratory tool, in which he questioned the lavish credit we give the scientist who is given patent rights to an idea, as opposed to the laboratory craftspeople who put in the long hours of actually making an experimental system function accurately and regularly, versus thirdly those who turn the tool into a marketable package that can be sold off the shelf (without special knowledge) and accrue the commercial value. Rabinow's account sounds not unlike Roy's case study above.

Roy's account of CERN's art residency program, however, sounds much like Rabinow's *Fieldwork in Morocco* book, oddly without a sense of what art-science collaborations can achieve. Part of the problem, perhaps, is the riddle: Is there an analogy, interdependence, or milieu of understanding between art and science? Many accounts of the arts and sciences at the fecund turn of the twentieth century note the crossover weltanschauung that was transforming from a mechanical universe to one recognizing probabilistic relationships, reflected in surface indeterminacies, inflected as well by mediating instruments whose capacities or affordances were changing and by observer-observed interactions. Some accounts of the arts and sciences today attempt to do something similar. Roy gestures at this, but with the unsatisfactory and empty term "postmodernism," a term from commercial branding efforts in architecture in reaction to overemphasis on streamlined functional aesthetics or, worse, from Alan Sokal's 1994 self-promoting silly hoax whose only stakes seemed to be whether he could fool an editorial board at a non-peer-reviewed journal to print his attack on STS as if it were a challenge to the doing of science. (I well remember the physicist Steven

Weinberg, who was being recruited by Sokal allies to attack STS, openly disappointing them by pithily saying that STS was not the enemy; the enemy was members of Congress who would not fund science appropriately. This was around the time of the defunding of the Superconducting Super Collider in East Texas and a growing sense within the senior physics community that their generation-long command of funding processes in Washington, DC, present since World War II, was coming to a close, not least because physics had been replaced by biology as the lead science for the coming century.)

* * *

Not all art-science initiatives generate results beyond the metaphorical or superficially pretty, but the best do. Some of these depend on technological innovation, such as Doc Edgerton's famous photographs of drops and splashes or bullets and bombs caught midaction by repeatable electronic flashes or, more recently, the extraordinary large-scale images of cellular processes that hang on the walls of the MIT Cancer Research Center, such as tracking nanoparticles across the blood-brain barrier, snapshots of tumor growth, or inspiration and respiration in a petri dish. Some art-science works do actual pedagogical work, such as the bioart of the Australians Ionatt Zurr and Oran Catts, in which they expand the work of the tissue-engineering lab into wider ecological realms, asking what kinds of social work are required to accomplish sustainability beyond an art installation's already-complicated importing of negative pressure hoods and other sterilizing protections. Or again, the physics department at Princeton a number of years ago gave a residency to an artist, and I helped bring him to MIT, who wanted to capture the sounds of the interior of a nuclear reactor: part of the project was to get the help of the machinists and material scientists to devise a recorder that would not melt in the reactor. The specifics of the project can be variously evaluated, but a key goal was to get scientists and engineers to work together in novel ways and see if they could learn new things. As to the dismissal of doing a walking tour through the tunnels at CERN, I can only counter that my visits to the tokomak and nuclear power plant at MIT made a profound impression that was useful years later in understanding the plume physics of the fires raging in the Western United States, Canada, Siberia, Australia, and elsewhere and that will increasingly become part of the new forms of scientific research on fire as the climate continues to change. This is critical work for surviving in the future, work that Adriana Petryna calls "horizon work." It is work that depends upon more than the predictability of old models and in that sense is a challenge to old science and demand for developing new science.

Much of art-science relations involves the innovation of visualization and now sonic technologies, a theme that I have been stressing in my teaching and exploring with colleagues like Joseph Dumit, who studies how scientists manipulate visualizations (and now movement) in order to see more clearly, or often just for publicity that requires the public to learn how to read visualizations for the information, not just the aesthetics. The covers of scientific journals are a rich source of visuals, but they often obscure as much in the name of prettiness as they reveal in the name of information. This, like coding, will become an increasingly important form of literacy.

Even if Roy's experience with CERN's artist-in-residence program was unsatisfactory, I urge readers to take the chapter as a placeholder or a challenge to look around their science and engineering environments for edifying initiatives and experiments. Experiments are just that: trials. Some work; many do not. Some work for some audiences; others work for other audiences. As science and engineering education becomes more and more project- and team-oriented, the artistic literacy side of creativity will grow. MIT's Media Lab stands as a powerful example of this sort of emerging pedagogy, as do similar endeavors at the Swiss ETH, Singapore's CREATE laboratories, and elsewhere.

REFERENCES

Beck, Ulrich. *Risk Society: Towards a New Modernity.* 1986. New York: Sage, 1992.

Biagioli, Mario, and Peter Galison. *Scientific Authorship: Credit and Intellectual Property in Science.* New York: Routledge, 2002.

Dumit, Joseph. *Picturing Personhood: Brainscans and Biomedical Identity.* Princeton, NJ: Princeton University Press, 2004.

Fischer, Michael M. J. *Emergent Forms of Life and the Anthropological Voice.* Durham, NC: Duke University Press, 2003.

——. *Probing Arts and Emergent Forms of Life.* Durham, NC: Duke University Press, 2023.

Hertz, Robert. *Death and the Right Hand.* Trans. Rodney and Claudia Needham. London: Cohen and West, 1960.

Fortun, Kim. *Advocacy After Bhopal: Environmentalism, Disaster, New Global Orders.* Chicago: University of Chicago Press, 2001.

Galison, Peter. *Image and Logic: A Material Culture of Microphysics.* Chicago: University of Chicago Press, 1997.

Kay, Lily. *Who Wrote the Book of Life: A History of the Genetic Code.* Stanford, CA: Stanford University Press, 2000.

Knorr-Cetina, Karin. *Epistemic Cultures: How the Sciences Make Knowledge.* Cambridge, MA: Harvard University Press, 1999.

Kondo, Dorine. *Crafting Selves: Power, Gender, and Discourses of Identity.* Chicago: University of Chicago Press, 1990.

Perrin, Constance. *Shouldering Risks: The Culture of Control in the Nuclear Industry.* Princeton, NJ: Princeton University Press, 2006.

Perrow, Charles. *Normal Accidents: Living with High Risk Technologies.* 2nd ed. Princeton, NJ: Princeton University Press, 1999.

Petryna, Adriana. *Horizon Work: At the Edges of Knowledge in an Age of Runaway Climate Change.* Princeton, NJ: Princeton University Press, 2022.

Rabinow, Paul. *French Modern: Norms and Forms of the Social Environment.* Chicago: University of Chicago Press, 1989

——. *Making PCR: A Story of Biotechnology.* Chicago: University of Chicago Press, 1996.

——. *Reflections on Fieldwork in Morocco.* Berkeley: University of California Press, 1977.

Rheinberger, Hans-Jörg. *A History of Epistemic Things: Synthesizing Proteins in the Test Tube.* Stanford, CA: Stanford University Press, 1997.

——. *Split and Spice: A Phenomenology of Experimentation.* Chicago: University of Chicago Press, 2023.

Traweek, Sharon. *Beamtimes and Lifetimes: The World of High Energy Physicists.* Rev. ed. Cambridge, MA: Harvard University Press, 1992.

Vaughn, Diane. *The Challenger Launch Decision: Risky Technology, Culture, and Deviance at NASA.* Enlarged edition. Chicago: University of Chicago Press, 2015.

ACKNOWLEDGMENTS

T his research project has been so long in the making that it is gratifying to acknowledge the few without whose support the book would never have been completed. James D. Wells worked zealously to save me and the project from dissolution. I feel blessed by the courage and vision he has shown in supporting me professionally when all the doors of postdoctoral sustenance had closed. I have profited incalculably from the brilliant and generous encouragement of Erik Mueggler. It was a stroke of sheer good luck to have come to his fastidious yet capacious attention. Thomas W. Laqueur shared with me his boundless enthusiasm for the underground dimension of European intellectual history, for which I am eternally beholden to him. Terrence Deacon provided unparalleled mentoring for close to two decades when others preferred collective amnesia to hard scientific questions. I should like to offer my deepest thanks to Lawrence Cohen for gently guiding me through the graduate program and beyond on a path that proved to be more punishing than either of us had anticipated. Stefania Pandolfo's willingness to discuss esoteric anthropological questions at any hour of the day or night made me feel like a colleague when I was a student. One of my biggest intellectual debts, and a very considerable one, is to César Gómez, whose arresting discussions on analytic philosophy and theoretical physics gave me the moral courage to persevere, without which every other help would have been of no avail. He also imparted the valuable lesson that it is better to be loved than to be admired. Luis Álvarez-Gaumé played the mischievous and endearing Mephisto to my Faustian project. Unstinting with his praise and delight for literary scholarship and critical and reprimanding of poor mathematical proficiency, which has made him a cherished friend over the years. Between Chander Mohan Bhatia and Punam Zutshi lies the arc of a dazzling intellectual nourishment of which the world knows nothing.

For the help rendered to me at CERN, I must single out Michael Doser, whose marvelous wit and lucid explanations helped me cruise through the entire duration of fieldwork. Convinced of the project's importance from the inchoate jumble I dished out, he provided me an office, which shows that the scientific field can receive adversaries with exquisite dignity. I benefited enormously from the insightful examples and explications of Albert de Roeck, which not only answered my questions but taught me how to raise foundational questions. Robert Fleischer persisted in discussing the basics of Standard Model physics and sympathetically listened to my research findings, whether or not he approved of my goals. Sudeshna Datta Cockerill and David Cockerill admitted me into their circle with so much warmth and so much understanding that it brightened my stay at CERN many times over. For the charming gift of her friendship and home, Monique Cahannes receives my everlasting gratitude. Little did I imagine when I stayed in her house on Rue Michel-Servet that I would taste the most wondrous experience of life. Had it not been for the lively spirit of the following physicists, engineers, and administrative staff in nurturing fellow travelers with unconventional interests, this work could not have come about: Francesco Bertinelli, Gautam Bhattacharyya, Fawzi Boudjema, Oliver Buchmueller, Cliff Burgess, Vinod Chohan, Alvaro de Rujula, Keith Dienes, Jonathan (John) Ellis, Lyndon (Lyn) Evans, David Francis, David Futyan, Pauline Gagnon, Fabiola Gianotti, Murdock (Gil) Gilchriese, Massimo Giovannini, Neal Hartman, Beate Heinemann, Ian Hinchliffe, Andreas Hoecker, Dilip Kumar Jana, Malcolm John, Gordon (Gordy) Kane, Wolfgang Lerche, Michelangelo Mangano, Thomas Mannel, Livio Mapelli, Claudia Marcelloni, Marcos Mariño-Beiras, Giuseppe Mornachi, Tatsuya Nakada, Satyanarayan Nandi, Holger Bech Nielsen, Markus Nordberg, Anne-Marie (Nanie) Perrin, Connie, Potter, Fernando Quevedo, Gijs de Rijk, Luigi (Gigi) Rolandi, Emma Sanders, Archana Sharma, Pietro Slavich, K. Sridhar, late Raymond Stora, Alireza Tavanfar, Tejinder (Jim) Singh Virdee, Folke Wallberg, Thorsten Wengler, and Daniel Wyler.

I would like to thank the University of Michigan, Ann Arbor, especially the Department of Physics, for graciously hosting me for four years, and the Department of Anthropology for the professional affiliation they gave me throughout the writing and production of this book. Continuous support has also been provided by the University of California, Berkeley, where Rosemary Joyce and Charles Hirschkind have helped me gain a slender foothold on the ledge of academic existence. The congenial working conditions at the Max Planck Institute for the Study of Religious and Ethnic Diversity, Gottingen, and the extraordinary goodwill of Peter van der Veer were indispensable to my attempts to round out this research. Always alert, accommodating, and chivalrous, Philippe Descola, at the Laboratoire d'Anthropologie Sociale, Paris, redefined my outlook of what a mentor can do for a junior scholar. I am immensely obliged to the Wenner-Gren

Foundation's Dissertation Fieldwork Grant, which enabled me to undertake initial fieldwork at CERN, and the award of the Hunt Postdoctoral Fellowship, which led me to complete the final version of the book. What I owe to Columbia University Press and Eric Schwartz for his outstanding editorial guidance and the careful reception of my ideas cannot be expressed in words. Indeed, I am thinking there must be a unique Greek word to describe the voluptuous editorial love that he brings to manuscripts. A special obligation must be acknowledged to Vitaly Pronskikh for goading me to think through the philosophical issues around parity violation. Michael M. J. Fischer thoroughly uplifted my spirits with his meticulous reading and incisive comments on the draft version, several of which I have incorporated into the text.

My greatest intellectual debt remains to my beloved teacher Jit Singh Uberoi. The spark he lit in my soul to give over to the unconditional love of truth and knowledge links my earliest university education to the present moment. Without understanding how it all conspired, I can only echo what Plotinus said after hearing Ammonios's first lecture: "This was the guide I was looking for." The influence of his monumental scholarship and discerning perspective exists as a palimpsest in every page hereafter. The book is dedicated to Jit Singh Uberoi and the physicists at CERN in whose company I spent many, many happy hours.

UNFINISHED NATURE

INTRODUCTION

T his book offers an ethnographic account of science in the making that delves deep into a particle physics laboratory to illustrate how some of its technical procedures and instruments are rooted in distinct intellectual assumptions about nature and society. I attempt to open up from within the modes of reasoning that animate the key subcultures of physics, namely, theory, experiment, and instrumentation, as revealed in the setting of a laboratory, which are decisive for the way in which a natural science understands itself and in its relation to the contemporary world. To be sure, this mode of contextualizing science, which tries to obtain an insight into the preconditions of its social existence, has long been available to us.[1] But what has been lost to sight is the elucidation of how a science like particle physics may incorporate elements into its domain beyond what its epistemic assumptions would lead us to expect, which deepens the mystery of what logic of classification it obeys. It is far from easy, however, to explicate the notion of classification, if only for the reason that it engenders notions of system, category, or context whose lucidity is hard to pinpoint in the scientific realm. That is to say, grave difficulties present themselves when we dig into what connects the cache of theories, calculative techniques, and experimental practices with the broader milieu in which a laboratory exists and that determines its access to funding, length of projects, and contact between research groups. So important a part have financial grants, political influence, and media access—that is, elements that compose the real world—come to play in the natural sciences that in recent years scholars have begun to suggest that the aggrandizing thrust of scientific knowledge is best understood as an empirical question.[2]

This is doubtlessly true, yet it must be admitted that one of the great puzzles for sociologists and philosophers is to explain how scientific discoveries and innovations constitute a break from a given empirical context. To any reader acquainted with modern particle physics, its most unusual feature is how it is constantly urged by its own ethos to go beyond the mere existent and through the emergence of new ideas and objects undergo a full transformation, which radically rearranges the stockpile of its concepts, theories, models, tools, and instruments. Experimental physics, especially, abounds with examples of novel phenomena created in the laboratory—from quarks to positrons—that show how troubling it is to maintain with consistency the meaning of the term *empirical reality*. Of course, a moment's reflection will also reveal that along with artifacts, which involve the fabrication of things into existence, experimental physics attaches equally great importance to facts that answer to true statements about how things are, which immediately introduces a rift in how we understand the creation of logical facts in relation to the creation of material objects. What sense would it make to say, writes Ian Hacking with caustic irony, that the universe began with a big bang is a fact that has been constructed?[3] Facts are not in the world in the same way as quarks and positrons are. Hacking has rightly drawn attention to the distinction of objects and ideas and why a seamless vision will not do for the difficult terrain of particle physics. It is impossible to overemphasize this fact. Baldly put, the gulf between experimental physics' proclaimed purity of facts with its prodigious capacity to create new artifacts underscores the extent to which empirical reality becomes problematic and devoid of specifiable meaning.

A fundamental premise of this book is that the context of discovery in particle physics cannot be explained by any compilation of empirical results because it expresses itself in the heuristic of conceptual possibilities. It could even be said that so long as we subscribe to empirical ways of thinking and seek an explanation of scientific knowledge in the strictly existent, we will never learn how particle accelerators deliver on discoveries, why signs and symbols dominate experimental physics, or what propels the search for grand unified theories. These questions have the widest and deepest relations not with any observable reality but with the immense power of intellectual classifications, which, as I will show later, create an ambiguous perception of what belongs to the milieu of modern social thought. Marx makes a penetrating observation regarding ancient art, which holds good for contemporary science: "But the difficulty does not consist in realizing that the Greek arts and epic are bound to certain social forms of development. The difficulty is that they still give us artistic pleasure and that, in a sense, they stand out as norms and as models that cannot be rivalled."[4] Marx's remark exposes the riddling problem of explaining how natural science is integrally bound up with the social milieu while it preserves its distinctiveness and promotes the expansion of knowledge, which furthers discovery, invention, and even experimentation.

In this book, the focus is the ongoing experiments at the site of the world's highest-energy accelerator, the Large Hadron Collider (LHC) at CERN, or the Conseil Européen pour la

Recherche Nucléaire, near Geneva in Switzerland, to unpack how concepts and classifications form precursors of scientific novelty. In March 2010, the LHC started its experimental run of proton-to-proton collisions as a probe into the structure of matter and forces of nature. The imposing character of these experiments was disclosed on July 4, 2012, when two of the collaborations, ATLAS and CMS, announced that they had discovered a new particle in the mass range of 125 gigaelectron volts (GeV) consistent with the Higgs boson predicted by the Standard Model.[5] Although the discovery of the Higgs particle was to some extent anticipated, it is not foretold what precise role it will play in the next few decades of the scientific era. Based on two and a half years of intensive anthropological fieldwork at the site of the LHC particle accelerator complex, this book follows in part the climactic discovery of the Higgs boson, which won the Nobel Prize in 2013, to tell the parallel story of what scientists have to say about their scientific commitments and concerns, the sources and vision guiding their experiments, and the questions they ask of themselves and of us in the social sciences. In this attempt, the narrative that emerges both imitates and rivals the exploratory stance of science to disclose not so much the discovery of a new fact as the discovery of new ways of thinking about a fact.

This is a site-specific ethnography and provides, in depth and salience, an account of the conditions of experimentation at the frontier of high-energy physics. However, in method and scope, the paramount focus is not of reportage or commemoration. A decade or two ago, it was fashionable among scholars of science studies to describe at length the zeitgeist of the laboratory they elected to study. But we have learned to be cautious after Tim Ingold's warning to the social sciences to be on their guard against ethnography's descriptive ideal.[6] Ingold's work is an important reminder that the vicissitudes of anthropological fieldwork among a community mean attending to people and things, learning from them, and reuniting knowledge and being, all of which lie beyond the pale of description. The situation is complicated in another way. The thrust of physics as well is not toward a description of the physical universe but an explanation into its inner constitution, which it undertakes by locating causes, discovering laws, and so on.[7] In that vein, it is not my intention to offer an exhaustive description of a science nor a prescription for a better science but to look closely at some of the presuppositions that serve in an interesting way to connect the technical procedures of a laboratory with wider principles of intellectual classification.

By presuppositions, I mean the class of beliefs that is collectively and unconsciously held by participants and of which they are unaware but that informs every aspect of scientific thinking and activity. Dwelling among the particle physics community at CERN, I observed that conceptions of matter and energy were derived from submerged assumptions about how the universe works. These assumptions frequently took the form of proscriptions and dualisms: human values do not affect physical nature; theory is separate from the domain of the instrumental or applied; conventions of measurement have no bearing on laws of physics; and so on. By piecing together how physicists deal with binary terms and relations, I gained a sense of their conceptual universe, and in the following pages, I aim to show how these beliefs, unnamed but presupposed, are present in physics as a living ontology. This is not to imply that they are somehow "given" in the datum of common experience or can be culled from individual utterances. Presuppositions, if I may try to make this concept a little clearer, occupy a space of reflection that may elude the consciousness of the individual scientist yet form a vital part of the collective scientific consciousness. This will be made abundantly clear by following a constellation of dualisms like those of fact and value, subject and object, and theory and practice, which make quite specific the link between humble laboratory operations, such as reading a histogram or computing the momentum of a particle, and more magisterial claims about the origin of the universe or the constitution of the atomic nucleus. Formulated in this way, it may sound a truism, but in fact, something is quite curious about the way in which the dualisms are organized, on which I shall have more to say in a moment. Properly speaking, the presuppositions belong neither to science nor to art. They are logical but also "practical," as Bourdieu uses the term, because they are kept up and cultivated.[8] That is to say, the presuppositions are integral to the work of physics. They justify themselves by being in constant use and by being the means through which day-to-day work as well as monumental events, such as discoveries, take shape.

Space will permit only a brief look at how the epistemic culture of physics digs into the roots of half-conscious beliefs that, parenthetically, betray their modern origins. In the theoretical sphere, the reigning paradigm of particle physics is called the Standard Model. It addresses three of the major fundamental forces—namely, electromagnetism, strong nuclear force, and the weak force—that govern interactions between elementary particles. While the Standard Model indicates that all physical reality subsists and exhausts itself within matter and forces, it is guided neither by matter nor forces but by the mathematical principle of symmetry. Symmetry denotes the property of invariance under certain classes of transformation, such as rotation, reflection, and translation. For instance, the premise of symmetry implies that an experiment performed in one location should have the same result as an identical experiment conducted elsewhere. That is, the result is said to be invariant under translation across space. Symmetry is such a compelling feature that physicists frequently use it to make predictions such as when matter is replaced by antimatter, the laws of nature will not change or if carrier particles exist for the weak force, which governs the decay of subatomic particles, carrier particles also must exist for the strong force, which is responsible for binding atomic nuclei. But merely expounding on new particles and force fields on the strength of symmetry considerations would not inspire confidence unless these are experimentally borne out. This is where the presupposition of nature having supreme sovereignty and physicists implicitly defending the freedom of nature in what it wishes to reveal during an experiment comes to the fore. Not surprisingly, we find that in the theoretical and experimental realm physicists are content to proclaim their passivity and simply intent on observing mechanisms and causes constitutive of physical nature. Attending to how such observations are made or contested lays bare the ideas of innovation and novelty and the more provocative question of understanding how possible facts obtain.[9]

Yet the matter of specificity of an experimental science like physics is not nearly as simple or straightforward as this. One of the most tantalizing unresolved questions in the Standard Model for the past fifty years has been how elementary particles acquire mass. Not all particles have masses. The photon, for instance, which carries electromagnetic force, has no mass at all. Why particles are not massless is a question that demands explanation. Here enters the physics of Higgs boson. As per the Standard Model, it is the interaction with the Higgs field that leads all particles of matter and forces, that is, fermions and bosons, to acquire a mass. In substance, particles are nothing but excitations of quantum fields that fill all of space. Because the mass of an elementary particle is a measure of its interaction with the Higgs field, the mass of the Higgs boson as such corresponds to the self-interaction of the Higgs field. But while the hypothesis of the Higgs field and the attending Higgs boson particle have been around since the 1960s, it was physically detected in experiments only as recently as 2012 at CERN. One of the reasons for the long wait is that the Higgs boson is an extremely rare and massive particle that decays rapidly and requires enormous energies to produce and detect it. Almost as soon as it is produced, a Higgs boson decays, taking less than 10^{-22} seconds. Its experimental detection can, therefore, be made only indirectly through the identification of the host of other particles produced in the wake of its decay, such as photons, leptons, and hadronic jets. The details of the detection of the Higgs boson from indirect searches for signals in the repeated collision and annihilation of hadron group of particles we must leave until chapter 2.

What is to be stressed here is that the production of a Higgs boson as a natural occurrence is a conviction that the physics community rarely abandons. Equally, the limitations of nature in producing some varieties of a stable Higgs boson are recognized on the plane of thought. Most physicists will insist the detection of a Higgs particle "is challenging, not in principle but for practical reasons."[10] Then efforts are made to find a way of overcoming these on the applied or engineering side. The outcome is satisfying because technology, propelled by human action, can achieve the energies and sensitivities needed to delve into subatomic distances and detect rare particles in head-on collisions. Thus, against the background of the developments in the Standard Model stands the technological innovation of the LHC, which can accelerate and search more particles emitted from the decay of atomic nuclei than any particle accelerator in the world. This shift between the belief that nature is supreme and the belief that nature can be subjugated—in practice—constitutes a key moment in experimental physics. Speaking more generally, the energy and intensity of colliding beams in the state-of-the-art LHC accelerator opens a window onto relations of pure and applied, thought and action, or fact and value, which draws out the centrality of the experiment at a level quite different from that which is empirically given. Moreover, these relations are neither arbitrary nor indeterminate but form a concatenated system.

To make sense of what an experiment in particle physics means, this study draws on the presuppositions that express, on the one hand, the opposition of subject and object and, on the other, the opposition of theory and practice. These two oppositions standing in the most intimate relationship, however, do not exhibit symmetrical relations to each other. The domain of theory is calculated to bring to light the operations of nature (object), in which the scientist (subject) carefully abstains from manifesting her presence and recedes to the background. But as soon as we pass to scientific practice, then the engineer (subject) is elevated, who with all possible ability and dexterity, proceeds toward the manipulation of nature (object) to transform it with a design and purpose in mind. To follow the double splinter of relations in forward and backward gears proves extraordinarily elusive. I will spell out more fully later how this concerted problem, in which a fundamental science makes it a matter of listening to the voice of nature by suppressing the voice of the scientist while technology makes nature plastic by intensifying the actions of the engineer, gives rise to endless confusion among critiques that boast of putting science in the dock. Important though it is to emphasize that without a thorough-going explication of the presuppositions, we cannot grasp the manner in which subatomic physics achieves its power, autonomy, and resilience. In sum, the narrative focus of the book is on multiple relations that dualisms of fact and value, subject and object, and theory and practice bear to each other to trace the conditions that have made the world's largest experiments possible at CERN. It is of capital importance to understand how a twenty-seven-kilometer underground accelerator that smashes two counterrotating proton beams every twenty-five nanoseconds at record energy of fourteen trillion electron volts, can dazzle in sheer physical intensity. But more alluring is the assurance and finality that eight thousand physicists and engineers affiliated with CERN bring to Bacon's famous simile of "putting nature on the rack."

My express purpose is to leverage the presuppositions to core concepts to extract a concession from physics on the full meaning and vitality of its discourse. It will be seen more plainly as we proceed that the power to make things real and active belongs to the domain of concepts.[11] Wherever we look, even for basic questions such as what are photons or how are neutrinos lacking in electric charge, we have to truck with concepts. Now, whether the concepts imply genuine essences, empirical abstractions, or probable certainties is a vast topic in itself, the scrutiny of which is beyond the scope of this book. However, the aspect that I want to draw attention to is that concepts in physics deal with properties that are considered self-subsisting or intrinsic to objects, such as electric charge, mass, or spin. Descartes had clamored, "extension in length, breadth and depth is what constitutes the nature of corporeal substance."[12] This seventeenth-century view crops up even today. The key thing required for identifying physical facts is the radical elimination of any form of subjective qualities like those pertaining to perceptions, feelings, and sensations (or sight, taste, sounds, and smells). For sure, physics cannot do without ears or eyes, but it is an entirely different

question whether these can be admitted in its cognitive inference, in its mode of knowing, or in the facts it arrives at. Certainly, among physicists, it is a matter of accepted wisdom that
they have successfully eliminated the human subject from their
concepts; in this regard, they frequently reminded me of the distance that separates the physical from the social sciences. This
may seem like a useless digression. But the point I am driving
at is that its concepts are solid enough to resist any attempt to
dilute them in an arbitrary or willful way. That is why I am prepared to give utmost importance to concepts and principles of
physics, no less than to its instruments or materials or artifacts.
This, though, is not the whole story.

Unquestionably, concepts and categories provide a clue to
the basic orientation of the scientific field. But a modality of
engagement like ethnography, which involves an intertwining of
lives, aims for something more. We are led to follow the meaning of concepts and principles not only in the interrelations they
bear to each other but also from the reception they are given in
the discourse and the counterresponses that ensue. Thus, the full
scope of science appears—outside of the success it achieves—in
what it signifies when it is raised to its own self-consciousness.
The dualisms, as emphasized earlier, are firmly entrenched in
the scientific mind and exert an abiding influence. In contrast, in
spite of its portentous capacity to enlist dualisms—and perhaps
more interesting—I found a few puzzling concepts in theoretical and experimental physics that confront and challenge, quite
effectively in my opinion, the divisions of mind and matter, fact
and value, or pure and applied in a discipline that ostensibly
proceeds from their strict separation. I believe this half of the
story on mediations to be more interesting given how they strain
against the boundaries of modern physics and reveal aspects
hitherto unnoticed by scholars in science studies.

In science studies the emphasis has largely been on unlocking the meaning of "big science" in everyday, situational features of a laboratory as a way to dismantle established assumptions and to shine the spotlight on local coordination and contingent outcomes in scientific production. The oft-repeated schematic of "disunity of science" amplifies that methodologies, inferential models, observational criteria, calculational tools, and experimental tests are so varied across the subcultures of physics that they preclude any vision of a unitary science.[13] The result is a critique that is proud of tilting the production of scientific knowledge to contingent, local factors. Karin Knorr-Cetina, for instance, vigorously maintains the significance of "the reversals practice brings about—reversal of universal standards through local conventions and opportunities, the reversal of rules through power, and the replacement of social characteristics of persons through situational features."[14] Admittedly, by questioning entrenched assumptions, we may disclose their provisional nature and raise the distinct suspicion that there can be other ways of structuring science. But unless the work of physicists worldwide is a series of blind happenings, we must ask, What is gained by surrendering to a rashness that remonstrates against physics as if it did not possess a specific character, thus undermining the compulsion that its observations, methods, and outcomes wield over us? More to the point, everyday events of a laboratory may not make apparent the logic or unity that underlies particle physics, but some shared presuppositions do exist or else we could not explain the enduring relations between different subcultures across laboratories, as I hope to establish in the book.

It would seem, taking a wide view, that earlier studies have not ignored the fact that at the edges of science, we are confronted with numerous instances of correspondence and mediation between natural kinds and human kinds. Scholars are unanimous that a bevy of mediating ontologies proliferate both within and beyond the walls of a laboratory.[15] At the same time, they are inclined to think that these unifying actions should be viewed more the result of a ruthless instrumentalism than any genuine harmonizing impulses. This sense of instrumental action is made prominent in terms like "hybrids," "assemblages," and "cyborgs."[16] It is also emphasized that the locus of mediations of natural and human kinds is to be found in the practices of science in which the action or doing component plays a singular role. Andrew Pickering believes that "the key advance made by science studies in the 1980s" is "the move towards studying scientific practice."[17] This may be a suggestive hypothesis, but any one-sided bid to discredit the legitimacy of factual claims put forward in the sciences on the material level of practices leaves unexamined the monumental role of thought or theory. Concepts in particle physics, like asymptotic freedom, which explains how interactions between particles get weaker as the energy scale increases, or models, like supersymmetry, which predicts the existence of shadow particles, are no less human for being more extravagant and surely merit reflection.[18] There is no reason we should be coy in approaching the technical concepts to discern features of subjectivity or contingency, and then we may be fully justified in regarding natural science as a normative activity or social construct like any other. Conversely, if we bifurcate and prioritize material practices over conceptual elements, the result not only produces a distorted view of physics and physicists but also obscures from view the tension that obtains between the domains of theory and practice. The relations between these two realms are anything but episodic.

To put the matter schematically, quite a big part of the problem of understanding science involves the coexistence of intrinsic (theories, concepts, epistemic beliefs) and extrinsic (political influence, worldly prestige, funding) aspects, which returns us to the relationship of science and society with which I began my observations. Speaking more generally, even among physicists, this is a topic of ardent concern. Several of them spoke to me about "the impact of science on society." The very idea bristles with difficulties. It assumes that the two domains are separated by a sharp break. Following Durkheim, David Bloor has shown in exemplary fashion that social milieu and scientific activity do not constitute two separate orders of existence in which the bone of contention is to know whether or not they should be set in association.[19] Rather, the two inflect and fold into each other as warp and woof. In this sense it may be said that physics *is* incorrigibly social: not because it is run by transnational collaborations or funded by taxpayers' money but because the social is in the structuring of its thought, in the persuasion of its instrumentation, in the conventions of its measurements, and in the efficacy of its concepts, areas from which I begin my own investigation.

Unyielding as such an approach might sound, readers should recall that the acid test for a framework that can go further in drawing out the limits of contemporary science is that it do so only without sliding into skepticism or relativism, which undercuts the possibility of any knowledge at all. Recent controversies, including the famous "science wars," have shown without doubt that the determining feature of modern science has been the rigorous assertion of objectivity both for its conceptual tenets, or facts, and experimental entities, or artifacts.[20] Furthermore, the two principal forms that objectivity takes—within limits—are of necessity and universality. That is to say, objectivity makes its appearance with the claim of probative force that everything it implies follows in accordance with the laws of nature, which are the same everywhere. Yet when we come to the chief objections against objectivity, these seem to be stacked disproportionately on the side of politics of knowledge, on murky interests and motivations of scientific workers, or on the liabilities of translations and inscriptions in the stabilization of knowledge claims. We have much to appreciate in how intricate networks of human and nonhuman interests and activities drive variegated practices of science, but I am not content with contextual or contingent associationism—and I phrase this cautiously—if only for the reason that these have not been effective against the challenge of objectivity in the core content of natural science. In particular, any attempt to interrogate the truth value of scientific claims whose determining ground is context and contingency is impotent because these are already outside the pale of universality and necessity. Therefore, it seems worth pondering, is there no language of relations with wider and deeper scope than assemblages or hybrids, which our current texts have accustomed us, to challenge the dualistic contentions of modern science?

Let me stress at the outset that the mediating links and correspondences perused here are not fortuitous juxtapositions tossed between extremes but relations that arise *necessarily* from the functioning together of concept and phenomenon. Along these lines, I have tried to seek that pivotal point at which opposed elements constitute a unity, which not only renders it distinct from the sum of its constituents but also makes it independent from anything external to it like context or circumstance. These two ideas should make explicit why a dialectical unity cannot be equated with more contemporary forms of assemblages, which have to do with combinatorial and contingent crossovers. To conflate an association of instrumental action with structural unity is, as I shall lay out in chapter 2, an error of oversimplification. Then again, a genuine synthesis of opposed tendencies has to have its locus in a form of thought specific to physics and not simply in segments of practices, which are many and diverse. In this respect, the concern is not so much with reality as with the epistemic foundations of reality and how we know them, which is the relationship to truth. As one would expect, the truth of a composite mediation is based upon the fact that we apprehend its force and substance in the phenomena themselves. And this is the larger point I wish to make—that by drawing together and establishing a nondual standpoint by which truths of nature may be phenomenally observed, we consciously engage with a different form of classification, which is not a pale reflection of modern science but insinuates a science that places the conception of nature squarely in relation to the human observer. We may, if we like, argue the matter the other way round: that any desire for achieving unity has to be grounded in a classification that locks into the common substrate of human and nonhuman nature.

As it happens, the empirical evidence around mediations in sociology or anthropology of science not only points to the contrast with dichotomous forms of reasoning but also attempts, more generally, a thinking-through of relations. These tendencies are not as mutually exclusive as they appear to be, though the latter is so wide that almost anything, from our DNA to set theory, may be expressed in the matrix of relations. To redress this, I draw on concrete and specific examples in particle physics that offer redolent lessons in relations that assume a nondualist form. One such concept is handedness, or the designation of right and left, to elementary particles as per the Standard Model, which is formulated as a chiral theory. The manner in which physicists classify particles such as "right-handed electrons" or "left-handed neutrinos" warrants the presence of a symbolic convention known as the "right-hand rule for angular momentum," where fingers of the right hand are used to indicate the direction of spin and orbital motion, which suggests to an extraordinary degree, even as a matter of metaphysical necessity, a human perspective to define the directions. If we step back and consider even the simple act of identifying a street as being on the left, or if we want to be seated on the right side at a table, we proceed with reference to the composite relations established from the *embodied* perspective of our immediate selves. Moreover, in some cases, the discernment of right and left involves a succession of other oppositions of pure and impure, high and low, and sky and earth, which turn on relations of cosmological proportions that serve as a primer to the complex sociological unity of the mind and the world. This part of the problem requires a lengthy discussion, and I will get to it in chapter 3, but it suffices to recall that in much the same way, particle interactions involved in electroweak interactions follow the distinction of right- and left-handedness, which exhibits the indissoluble unity of the human observer, social conventions, and the physical universe. What readers can glimpse is the astonishing expansion of a concept of right-left orientation that communicates the presence of relations not as emergent from a plurality of distinct elements but as necessarily blended in the sense of a unity in a manifold.

It is instructive to consider a second specific example, of signature, which sets forth phenomena of nature manifestly coeval with the scientific subject. In experimental physics, searches for new particles are often conducted by looking at signatures, or the final decay products emerging from the debris of high luminosity in high-energy collisions. But what constitutes a signature event, such as extrapolating the creation of a Higgs boson from a pair of high-momentum photons, requires the exertion of human judgment, which straightaway collapses the dualism of fact and value. The very abstraction of photons in the debris of collisions as a signature event denotes a judging mind, which belies the claim of pure objectivity on which particle physics thrives. I have kept the discussion to a general level, namely, the manner in which scientists conceptualize the existence of new particles requires that the two strands of fact and value reinforce each other. However, an unfathomable depth runs through the signatures of physics, which betray a connecting link with the terrain of semiotics or semiology and which obliges us to recognize that physical reality demands a grounding in the symbolic plane. This is paradoxically brought out when we concentrate on radical or unanticipated discoveries, such as the idea of exotic particles, where we are faced with the curious case of signs existing for things that might not exist. This liquidation of ontological priority in the case of unknown or new particles not only creates complications for die-hard materialists but also demonstrates that relations of symbolism, language, and reality are set in mutual concordance and not assembled piecemeal in additive fashion.

It is no exaggeration to claim that the relationship of perspective and judgment, which the two concepts of signature and handedness express, enacts a refusal of objectivity from the heart of physics. The evidence of ethnography testifies to this. Scores of conversations with physicists on how the constitution of signature events of the Higgs boson evince their judgment or why the polarity of right- and left-hand in electroweak interactions reflects a human perspective helped bring to surface their astonishment at what are flatly held to be "statements of fact." From the start, the book focuses squarely on capturing their way of viewing things especially around the movement of these nondualist, nonmodern concepts, which is of significance because most physicists are, if I may spell it out, neither indifferent nor unyielding to evidence that is critical, well founded, and intrinsic to the content of their science. Speaking in a more general fashion, the book adopts a standpoint that follows through the necessary interconnections of things, which provides a lasting foundation to refute the claim of scientific objectivity.

Yet more than donning the mantle of subversion, these concepts are absorbing to the extent that these are persuasions grounded in truth. In this respect, debates in the field have had a profound influence on the structure and substance of arguments presented here, and I make no secret of my debt to the insights gained with interlocutors during fieldwork. I recall that in the initial months at CERN, I was trying to schedule an interview with Marcos Mariño-Beiras, a theoretical physicist, who seemed somewhat reluctant to commit to a day and time. A few months later, I chanced upon him in the corridor of the Theory building and detained him. In the course of the conversation Mariño-Beiras said to me with bitter irony that anthropologists revere the utterances of Indigenous Americans or Australian Aborigines of openly animistic import, but physicists' assertions on the objectivity of their science are taken to be empty rhetoric. The significance of what he said was so great for ethnographic purposes because until then I had not thought of objectivity as a defining trait, and as I came to reflect on his incisive observation, I knew my immediate task had to be a grasp of their technical vocabulary. Even so, more than the sense of injury at not being taken at face value, the *one* refrain of physicists was whether anyone could name a concept of physics tinged with subjectivism. The question was not entirely rhetorical. I tried to point to the use of metaphors and analogies in their discipline, which reveals interesting interconnections of fact and fiction, but they cringed with offence on the grounds that literary devices are incidental to ontological claims. Similarly, hybrid or contingent associations found in the realm of practices were thought to be too meager and commonplace for a complete takedown of physics.

We must not overstate matters: distinguished attempts from within and outside science have questioned the manner in which "pure" facts are distilled by means of mechanical observation and quantitative enumeration. In the teeth of all opposition, Gottfried Leibniz showed that material bodies possess a "living force" or *vis viva*, whose measure is given by mv^2, which suggests that besides extension and motion, substances are capable of activity, resistance, and force that defies the bounds of mechanical reasoning. Johann Wolfgang von Goethe went to his grave insisting that in optics, the unilinear prismatic spectrum of the Newtonians was deficient for ignoring the contribution of the human eye to the production of colors. In the light of these historical precedents, this book strives for a concrete assessment of concepts like handedness or signatures, which are, properly speaking, invocations of the unity of fact and value, object and the subject, that modern physics has all but suppressed, both in practice and in principle.[21] In anticipation of what will emerge fully later, I take the opportunity to underline that reflecting on the interconnections between the human and the nonhuman that recall their necessary unity, as I term it, opens the door to a genuine, precise, and profitable postconstructivist appraisal of science. The intention of this remark is purely to emphasize that the book is concerned with narrowing the gap in the mutual understanding of the social and the physical sciences. That is the only way forward from the storm and the stress of "two cultures," eloquently baptized by C. P. Snow and still smoldering in our times.

Finally, to forestall any misunderstanding, I should clarify that while it is true that concepts like signature or handedness cannot be satisfactorily explained if we leave out of the account the premodern classifications out of which they arise and with which they remain closely connected, I will not drift into analyzing whether these are survivals or residues of a remote age. These would be second-order, speculative questions. The point is that modern science has a prodigiously expressive character but is rarely considered as an art of expression. Only when we step into the living conversations of science, even the most quotidian and rudimentary, we confront the hidden synthesis of its manifold practices and abstractions. In pursuit of this aim, this study draws on a range of sources from contemporary science and technology studies scholarship to the history of science, philosophy, semiotics, and especially anthropology. Throughout the text, real names of informants have been used after obtaining their consent. I should add that the rich harvest of notes and observations gained through anthropological fieldwork, comprising close interactions with more than a hundred physicists and engineers, was in part aided by the fact that I was officially affiliated with the ATLAS experiment at CERN during fieldwork and afterward. All things considered, what remains remarkable for us who come from an ethnographic immersion in particle physics are the ways in which its conceptual trajectories disclose how the culture of modern science is also one of forgetfulness and why it presents its own unity as a particular kind of puzzle or problem.

Now a word on the organization of the book, which includes five chapters. Each chapter centers around a particular tension that conceptualizes what is implicit to the epistemic culture of particle physics and constitutes a sufficient whole. The material and methodological emphasis is on the setting up and eventual grasp of specific, concrete problems. Chapter 1 begins by putting in perspective the search for the Higgs boson on the experiments at the LHC. As I said, the LHC is an accelerator that collides two counterrotating beams of protons head-on with each other every twenty-five nanoseconds at fourteen trillion electron volts of energy. Particle acceleration and collision is based on the principle of mass-energy conversion ($E = mc^2$) in which fast-moving particles are collided with each other so that some of their energy is converted into the creation of new particles. Four major experiments, dispersed around the twenty-seven-kilometer circular accelerator, study the data ensuing from the collisions with the aid of massive detectors situated in underground caverns. It was the combined data of two of the largest experiments, ATLAS and CMS, that confirmed the discovery of the Higgs boson in July 2012. The chapter capitalizes on the Higgs boson breakthrough to make two points: First, the worldview of physics feeds on the theme of constant innovations in ideas and techniques, which defines the extent to which symbolic elements of prestige and influence gain ascendancy over purely pragmatic or instrumental aspects of scientific life. Second, while the relation to technology is the most consequential to the production of discoveries and innovations, how the material and the conceptual aspects join together calls for an examination of the underlying logic of contemporary physics in its entirety.

In chapter 2, the opposition of fact and value, or thing and sign, is explored in the concept of signature commonly used in experimental physics. The detection of new particles in experiments is a challenging task. There is something specially rarified about the Higgs boson for it can never be observed directly in particle detectors. It is an unstable particle that decays rapidly into lighter particles after being produced in collisions. In this respect, we should appreciate the importance of a Higgs particle decaying directly into two photons, a signature termed as the "golden channel" in the physics community, as it is relatively easy to discern. At the same time, photons in the detector are emerging from *bremsstrahlung*, or electromagnetic radiation given off by accelerated particles when forced to orbit in a circular trajectory.[22] These photons constitute the background. Accordingly, an examination of the source of the emitted photons has to be made, which implies an evaluation in which physicists have to exercise their judgment to distinguish signal photons from background photon events. In this chapter, I establish how the sheer act of identifying a signature event closes the gap between fact and value. In the process, I also highlight some of the semiotic features of relations between signs and things, which opens a window into the medieval "doctrine of signatures" more generally.

A new problematic emerges in chapter 3. This time, I explore the radical opposition of subject and object through the core concept of chirality, or handedness, to raise the question: Does physics admit of orientation? The notion of orientation, that is, designating right and left in space, acquires meaning by virtue of a perspective (of an observer) and certain conventions (of the community). But if physics proceeds from the separation of subject and object, then how can it posit—as it does—particles with a preferred orientation? This question drives me to revisit the work of the French anthropologist Robert Hertz, who ingeniously recognized that the act of differentiating between right and left cannot exist as a pure abstraction; it is a concept centered on our embodied, living, thinking selves. I then try to show how the concept of chiral orientation is captured in the quantum mechanical property of spin, or intrinsic rotational motion of particles, which defines them as being "left-handed" or "right-handed." A remarkable feature of electroweak interactions is that nature almost always shows a slight preference for left-handed particles like electrons, muons, tau leptons, and neutrinos over right-handed particles. Following through the asymmetry of right and left, dubbed *parity violation*, gives us a glimpse of how social conventions governing particle orientation operate in relativistic quantum physics. Here too the discovery of the Higgs boson acts as a midwife to richer sources of understanding the asymmetry of right and left. But what prepares us for something even more daunting is that measured against its own ideal of objectivity, the concept of chiral orientation curiously appears to move physics backward—toward the subject.

In chapter 4, the focus shifts to the material culture of the laboratory. After all, the accelerator forms the real pride and arrogance of high-energy physics.[23] The technology of the LHC is novel. For collisions it uses hadrons, such as protons, which are composite particles. Some 1,600 superconducting magnets encircle the beams to bend and deflect the protons for acceleration. To keep the electromagnets in this superconducting state, CERN's engineers have designed a cryostat facility that maintains a temperature of 1.9 kelvins (−271.3°C) which, as one of its posters proclaims, makes "LHC the coolest place on earth!"[24] However, an explosion occurred in the LHC tunnel on September 19, 2008, which brought the epic collider to a standstill. This incident provided an extraordinarily rich opportunity to investigate the relation of temporality to the cycle of work and grasp the division of labor obtaining in the laboratory. In chapter 4, I consider how the crisis following the accident discloses a classification of physics at the heart of which lies the opposition of mental labor and manual labor, which also provides satisfactory answers to questions involving the interface of experimental physics and engineering skill.

Chapter 5 turns to what is perhaps the most unorthodox experiment undertaken by CERN so far: seeking an engagement with art. Beginning in 2009, CERN launched a "cultural policy" to invite professional artists to spend a few months under its roof with the desire to achieve higher intellectual synthesis as well as shrewdly uphold its ascendancy in producing universal or capacious truths. At the core, the policy initiative involving the arts engenders the view that emotions and spontaneity are not excluded from the laboratory's focus and are welcomed alongside reason and systematicity. Readers will recall that a few years ago Alan Sokal, a mathematical physicist, played a hoax to single out postmodern literature for its circuitous jargon, slipshod standards, and disregard of disciplinary boundaries. Sokal's hoax will be considered for analysis to argue that postmodern tendencies are not confined to the humanities, and the tables seem to have strangely turned when we find a physics laboratory preaching to us with gusto on the stirrings of aesthetics and emotions using high-flown jargon. The chapter is introduced by an examination of CERN's "cultural policy," followed by an account of why the organization's higher management considers that consorting with the arts boosts the prospects of particle physics. It concludes with an analysis of how the adversarial and fraternal relationship of art and science fits with the logic of mélange that postmodernism favors.

A few remarks about the field site are in order. CERN is a particle accelerator complex that straddles the French-Swiss border and is situated about eight kilometers from the city of Geneva in Switzerland. As an international facility, it is exclusively neither under Swiss nor French jurisdiction and is privileged with the status of diplomatic immunity. CERN officially came into existence in September 1954, when representatives of eleven European nations ratified the organization's resolution toward developing and sharing the costs of a nuclear physics research facility. The exodus of scientific talent westward to the United States at the end of the Second World War had led to tremendous soul-searching on how to create an enduring global platform that would secure Europe's lead in science and technology in the future. The effort's culmination was CERN. The memory of exodus and impoverishment in the postwar situation is one of several reasons why financial contributions of CERN's member states toward the organization's budget are pledged well in advance. As a result, CERN can often benefit from temporary financial borrowings for a fixed number of years, if it has to, to meet the costs of construction of colossal projects, like the LHC. This is unlike the situation in the United States, where science budgets often come under strain from changes in political administration. Apart from the European member states, six countries enjoy "observer" status at CERN, which include India, Israel, Japan, Russia, Turkey, and the United States. One could say that CERN is a European laboratory hosting a global project.

The total number of personnel employed by CERN is 2,250, out of which approximately 20 are from the Theory Division; 50 are "research physicists," that is, experimentalists who are tasked with data analysis; 300 are "applied physicists," who deal with research and development, construction, installation, and commissioning of the detectors; and over a thousand are technicians, mechanics, and engineers involved with accelerator and beam instrumentation. The bulk of the workforce, however, does not come not from CERN directly but from worldwide universities and research centers, which are participating on the various experiments at CERN. Some 9,000 scientists, representing 580 universities and 85 nationalities flock here for their research. That numbers is a pressing issue at CERN is obvious from the constant expansions that are carried out by the organization in parking spaces, office areas, and cafeterias. Although in appearance it is an academic establishment—with most people who pass through its corridors holding doctorates—it has the underpinnings of a global corporation with markers of worldly power and success.

The head of the CERN management is the director-general, whose term lasts for five years. The "CERN Council," comprising the European member state, is the highest decision-making authority of the organization. The council is responsible for all of CERN's activities in scientific, technical, and administrative matters and is assisted by two committees, the Scientific Policy Committee, which evaluates the scientific merit of activities proposed by the physicists, and the Finance Committee, which deals with issues relating to budget and expenditure. The CERN Council meets periodically, and in particular some three to four times a year as a "Strategy Session," to formulate a comprehensive plan on how to harness resources for the European laboratory. The physicists often make jokes and witty puns on these strategy objectives, which we can skip. The LHC is the current flagship project of CERN. It is expected that for the next two or three decades, it will be the sole dominant player in experimental physics globally. Some of the key questions that the LHC experiments are poised to tackle relate to the origin of mass, the nature of quark-gluon plasma in the early universe, the asymmetry between matter and antimatter, and the existence of extra dimensions of space-time, among others.

CERN is a pure science laboratory. Physicists frequently quip how hard it is for them to justify to funding agencies and the public the "spin-offs," or the benefits of particle physics for applications such as in laser technology or in medical imaging. In conversations, nearly every one of them affirmed that they could have gone into more lucrative professions, such as banking, finance, engineering, or architecture, but that their ambition is of another sort. They find a distinct resonance with Max Weber's depiction of science as a "calling" rather than as merely a profession.[25] Most physicists can be described as workaholics. They are also zealously peripatetic. I experienced and strategically effected much of my electronic communication while they waited to fly in or out of international conferences. Social events like staff picnics and conference banquets are periodically organized within the set circles of every subculture, theory, experiment, and instrumentation. However, the Christmas dinner, annually hosted by CERN's Theory Division, does not follow this compulsion and draws together a richer assembly of staff members. Physics dominates the conversations in the ATLAS control room as well as lunch or coffee meetings, but these are also conducted in the higher sense with the belief that they are striving after grander truths. Nonetheless, most physicists cannot fully rid themselves of mechanical and atomistic reasonings. To these general and certainly inadequate reflections, I should perhaps add that most vital is their conceit of mathematics. What proved most vexatious in the early days of fieldwork was that coming from the pure social sciences, I was not considered equal to attending to the intricacies of their discipline. It was sufficient for me that they did not begrudge spending enthusiastic amounts of time explaining the high points of their métier in the nuances of complex numbers and matrix multiplication. By the time fieldwork was completed, a few of the informants extolled how the anthropological encounter had introduced a lively intellectual stimulus in their circle and, to some extent, produced a feeling of esteem for a science other than their own. They even argued that it amounted to breaking the impasse of the "two cultures." How far this is true is a reflection best left for another occasion.

To sum up, as a location of a fundamental science and in its capacity to address key questions relating to the physical universe, CERN attracts a great deal of press coverage, which the physicists seem to relish. Given at once to relatively esoteric concerns of particle cosmology, they are equally savvy with media blitz and

CD-plated cars.[26] They consider themselves to be endowed with real powers of invention as well as possessing adroitness in tackling difficult cosmological questions and arranging the reflections in a manner that makes them plausible to the educated public. Recognition and reward, like the coveted Nobel Prize, is highly desired and sought after, resulting in a ruthlessly competitive milieu at times. The spirit of competition sits uneasily with the demand for large-scale collaboration that contemporary experimental physics makes. This tension, between competition and collaboration, while demoralizing at an individual level, is overall considered to be healthy for the field. Like any other community, the particle physics community exudes a strong sense of self-importance in the universe's scheme of things. The stirring belief in one's importance is crucial to the narrative and action of what takes place at CERN and, among other reasons, makes it an exemplary site for an ethnographic study.

1

FINDING THE HIGGS BOSON

Necessary truths must have principles whose proof does not depend upon the testimony of the senses, although without the senses it would never have occurred to us to think about them.
—Gottfried Leibniz

I n March 2010, the Large Hadron Collider (LHC) started its run of proton-to-proton collisions to search for the last remaining unobserved particle belonging to the Standard Model, the Higgs boson (figure 1.1). The Standard Model is a theoretical explanation of the fundamental forces interacting between elementary particles. Aside from leaving out gravity, it "covers everything else, from the reactions that fuel the sun to the forces that hold a snowflake together,"[1] and is hailed as the crowning achievement of twentieth-century particle physics. It is noteworthy that the Standard Model had postulated the existence of a spin-zero, or scalar particle, like the Higgs boson in the 1960s. In 1964, Peter Higgs, among others, had proposed that a field and a particle associated with this field pervade all space and are responsible for giving particles masses. The actual particle, however, had eluded experimental detection.

FIGURE 1.1 Map of the area occupied by the LHC at CERN.

This is not surprising given that it is a rare particle that is produced only occasionally in collisions—about once in a billion times. It is extraordinarily difficult to identify in the debris of colliding particles because it decays in the tiniest fraction of a second into lighter particles. Regular experimental searches for this particle have been carried out, starting with the Large Electron Positron (LEP) collider at CERN, which was the immediate predecessor to the LHC, as well as on the Tevatron collider housed in Fermilab, Illinois, but to no avail. "Now the most urgent question in particle physics (maybe in physics as a whole) is: where is the Higgs?"[2]

I arrived at CERN for fieldwork in August 2007. In the beginning, I spent several weeks interviewing physicists and asking whether they looked forward to the discovery of the Higgs and the final validation of the Standard Model. To my surprise, they seemed gloomy at the prospect of its discovery. My first meeting with Luis Álvarez-Gaumé, head of the Theoretical Physics Division at CERN, was an eye opener. With great excitement, I asked him about the imminent possibility of a Higgs discovery on the LHC. He replied, "No, no we do not want to see the Higgs. The field will be totally dead. The press office has created this impression that the LHC has to do with the search for the Higgs. It is most unfortunate. We wish to see new particles, not the Higgs boson." He went on to observe that physics is an experimental science. The Standard Model had codified the understanding of the physical universe in a powerful paradigm, but the paradigm was exhausted. Finding the last remaining particle might confirm Standard Model's predictive power, but at the same time, it would offer no new directions to particle physics. I was struck by his reply. A few weeks later, I met John Ellis, a CERN staff theoretical physicist, and spoke to him about the LHC's extraordinary energy range in detecting the elusive Higgs particle. Ellis visibly shrank from the boldness of the experimental venture and conveyed with great moderation and professional exactitude that their community does not care about finding the Higgs particle. He said they are more interested in exploring new and unknown frontiers in physics. I demurred and asked whether any physicist had written what the community thought. He immediately drew my attention to one of his articles published in *Nature*, in which he had penned without reservation, "Theorists are amusing themselves discussing which would be worse: to discover a Higgs boson with exactly the properties predicted in the standard model or to discover that there is no Higgs boson. . . . The absence of a Higgs boson would be exciting for particle physicists, but it might not be so funny to explain to the politicians who have funded the LHC mainly to discover this particle."[3] I was baffled on reading it. The contrast between what the media or popular science books were proposing and what the physicists were articulating could not be starker. While the search for the Higgs particle seemed to scream from every poster adorning the corridors of CERN, the community appeared lackadaisical. The physicists were more enthusiastic about hints of physics "beyond the Standard Model," also known as "New Physics," in the LHC collisions.

Of course, the community has compelling reasons to probe a future of physics beyond the Standard Model. In spite of the fact that all experimental tests have agreed fully with Standard Model predictions—and this is true to date—the model has a number of unresolved issues. It lacks robustness at higher energies and is only an "effective" theory at low-energy scales. It has a large number of arbitrary elements, called "free parameters," which are entered by hand.[4] The Standard Model also does not offer any explanation on differing masses, such as, why is the top quark, the heaviest-known elementary particle, around three hundred thousand times heavier than the electron, another elementary particle.[5] It does not account for the observed phenomenon of "neutrino oscillations," which suggests that neutrinos may possess a small but finite mass and cannot be reconciled with the framework of the Standard Model. That is, any explanation of neutrino oscillations has to be, by default, under the purview of New Physics of some sort. All such open questions compel the community's efforts to explore the possibilities of New Physics beyond the Standard Model. But it was not just talk of New Physics that was rampant in the initial weeks of fieldwork. The concern for novelty was apparent in a few of the proposed instrumentation projects. Work was already underway on the next set of accelerators, like the International Linear Collider (ILC) and the Compact Linear Collider (CLIC). Because the timescale for building accelerators is so large, the research and development (R&D) for rival projects like the ILC and the CLIC had commenced, and the design and prototyping stages were already completed. When I heard this in 2007, I was amused. At the time, the LHC had not even started yet everyone showed unmistakable signs of enthusiasm in the prospects of new discoveries that would exceed the Standard Model. The question before me was that if the Standard Model works and works fairly well, then, what motivates the physics community to go beyond it?

"We have never found any deviation from the Standard Model, so we have been right for 30 years. It is sort of boring," a senior theorist at CERN, Alvaro de Rújula, reflected candidly in a conversation. de Rújula was not alone in citing boredom with standard (model) physics. When I asked César Gómez, "Why do you work on string theory?," "I don't know how else to keep myself entertained" came the prompt reply. Not the response I was expecting, but whether uttered in jest or in earnest, we should not disregard physicists' statements on boredom or ennui as motivating factors that propel scientific research. The physics community's emphasis on novelty is highly illustrative of scientific work. The eagerness for new particles conveys that an expressive or a symbolic aspect, along with a purely instrumental aspect, is pertinent to discoveries. Their manifest lack of interest in the Higgs boson means that innovations in science do not always proceed from a sense of failure or a desire to meet fixed objectives but from an unending quest for novelty.[6] We are encouraged in this line of thinking from an observation made in the sphere of commercial technology, in which innovativeness reaches such a climax that "the rate of obsolescence overtakes the rate of depreciation." That is to say, ideas and products are phased out even before they get worn out.[7] Uberoi argues that what characterizes techno-science in modern times is not simply an exponentially fast rate of growth but rather the abandonment of things before they have run the course of their usefulness, because they are "no longer considered fashionable or acceptable as up-to-date."[8] The force of this argument is closely tied to the utterances of the physics community on the desire to script a new future even when the contemporary state of affairs is a fairly satisfactory and successful one. In this quest for novelty, physics betrays that its goal is not that of finality but of development and growth per se. Although consensus has not been reached on the prognosis of the future that will emerge, the trend is seen as desirable, healthy, and normal. According to the ATLAS technical proposal, "We have before us a situation in which the existence of new physics has been identified within a defined framework, but the answer itself is quite open. Such a circumstance is very attractive in targeting our scientific efforts."[9] The open and contingent future is a thrilling prospect and an indication that all is well. Therefore, it is not entirely for the sake of a more complete picture that New Physics prospects are being contemplated. In fact, the possibility that those may even upend and undermine the Standard Model is readily admitted and not shied away from. The conclusion that we may draw is that while Standard Model physics is *important*, it is no longer *interesting*.

Max Weber had eloquently recognized that it is in the nature of science for accomplishments to become antiquated. He wrote,

"Every scientific 'fulfillment' raises new 'questions'; it *asks* to be 'surpassed' and outdated."[10] Weber's point was that far from being extreme, which is how some of the informants at CERN delighted in presenting it, this aspect of techno-scientific life is symptomatic of modern science. Looking to render one's ideas obsolete is not an exception or an aberration for Weber. It simply points to the chasm of fact and value that is achieved under modern life. The ability to think or act without regard to ultimate values is perfectly normal for modernity. Weber should not be read as rejecting science, rather he is rejecting science as a source of value claims. Irrational as it may sound, the quest for novelty for its own sake is not illicit, but it is what defines progress. In this spirit, it is typical that science should exceed its limits, and every so often, it does so. The spirit of relentless discovery and innovation in the natural sciences is best understood in Feynman's quip that doing "physics is fun." He wrote, "For me, physics is more fun than anything else or I couldn't be doing it."[11] Without a doubt, a definite norm is present in the light of which physicists' expressions of excitement or ennui makes sense.

What about the opposite? Does particle physics involve practical efficacy? An article in the *Guardian* titled "Is the Large Hadron Collider Worth Its Massive Price Tag?" clamors: "what price society is willing to pay to understand the universe."[12] The LHC has been built at a staggering cost of $6.4 billion. The United States has contributed approximately $521 million and the rest has been borne by the twenty member states of CERN. A steady stream of critical voices in the media have asked why billions must be spent to collide two tiny atoms in the hope of discovering a new particle. In other words, should we not do something more useful with the sum of money? In February 2008, the Science and Technology Facilities Council (STFC), which funds all of the United Kingdom's particle physics and astronomy research, announced massive budget cuts to the Large Hadron Collider unless it could give convincing reasons on why particle accelerators are productive: "In times of global financial meltdown and looming environmental problems, it's not unfair to wonder whether this kind of basic research is a luxury we can't afford. It's a question the physicists ponder and perhaps never fully answer."[13] This remark should clarify that innovations in physics rarely resolve practical problems in any direct way. An anecdote may illustrate the point better. When Robert (Bob) Wilson, the first director of Fermi National Accelerator Laboratory at Batavia, Illinois, was asked by a Congressional Committee, "What will your lab contribute to the defense of the [United States]?" he replied "Nothing, but it will make it worth defending."[14] Without doubt, the overall power and prestige of high-energy physics derives from the successes of the Manhattan project and the Cold War,[15] today's physics has no direct political or military benefits to offer. In CERN, the spin-offs of particle physics for industry, commerce, and technology are touted only lackadaisically. It is understood that the Large Hadron Collider project stands for an extraordinarily ambitious intellectual adventure rather than for any instrumental benefit. According to a CERN staff theoretical physicist, Gian Giudice, "The most fascinating aspect of the LHC is its journey towards the unknown . . . which is carried out with complex cutting edge technologies and guided by theoretical speculations whose understanding requires knowledge of advanced physics and mathematics. These are the very aspects that have shrouded the work of physicists in a cloud of esoteric mystery, discouraging the interest of the uninitiated."[16]

This quote is from his popular book, *A Zeptospace Odyssey*, which introduces LHC physics to lay audiences using nontechnical vocabulary. Giudice skillfully makes the argument that pure science does not serve a greater purpose or end, and as it feels its way toward the heart of things, which quest never relinquishes, it also surpasses the answers it finds in search of new questions and problems. Thus, while the Higgs hogs the popular imagination, physicists are terrified that the Higgs particle would show up with no other revelations, thus concluding the Standard Model paradigm of particle physics. If physics today delivers on all the goals of explanation, it would be dead. To be alive and in business in the present, it must constantly create futures that can later be surpassed. However, this sense of unlimited possibilities—their spontaneity and their sweep—does not lead us to a lurking absurdity because the physical sciences tend to yield constant and sufficient objects, including everything from electrons to black holes. This sense of unlimited possibilities also does not imply that anything goes. We are certainly far from the suggestion that the spirit of innovativeness that dominates physics is one of unbridled imagination. The point is succinct: the logic of discoveries is tethered to possibilities. Now we must ask, how does this emphasis on constant novelty square with the ground situation?

THE UNIT OF INQUIRY

Notwithstanding all of the talk around innovations and futures, particle physics has been in the grip of long-standing inertia. A stark contradiction exists between what the physicists seek and the rut prevailing in the field. I was told with pedantic exactitude that the decades leading up to the establishment of the Standard Model were stimulating for the particle physics community, both theoretically and experimentally, with significant breakthroughs

made one after another, such as the discovery of three generations of quark and lepton pairs composed of matter, asymptotic
freedom in strong interactions, and neutral currents that led to
the unification of the "electromagnetic" and the "weak" force,
among others. After this "golden age of modern particle physics,"[17] the field sputtered and entered stagnation. According to
Steven Weinberg, Nobel laureate and a towering figure in particle physics, "Since the late '70s, I'd say, particle physics has been
in somewhat of a doldrums."[18] Since the 1970s, every new experimental finding has corroborated Standard Model predictions
like clockwork, including the discovery of W and Z bosons on
the last major experiment at CERN in 1983, and the experimental validation of the Top quark in 1995 at the Tevatron, Fermi
National Accelerator Laboratory, Illinois.

The enduring problem of contemporary physics is not only
that new experimental findings and data have been lacking
but also, in fact, that hardly any new data have been reported
because of the slow pace of development of accelerators after the
decline of the Cold War.[19] As a result, experimental physicists
have been forced to work with Monte Carlo, or computer, simulations rather than with "real data" of particle collisions from
experiments. David Francis, who was the Trigger/DAQ project leader of the ATLAS experiment when I met him, wistfully
observed that fifty years from now, historians of science looking
at this period would find only a plethora of Monte Carlo plots
and histograms. The lack of real data is a hurdle for budding
careers in experimental physics in particular. For instance, in the
United States, doctorates in experimental physics are, as a rule,
not awarded to work done on the basis of Monte Carlo simulations but instead require the evidence of real-time collision data.
An article titled, "LHC Students Face Data Drought," appearing in *Nature*, neatly encapsulates the frustration and anxiety

of students in experimental physics with respect to their future career prospects and the need for accelerators and experiments.[20]

The LHC was originally conceived in 1983 and was approved for construction by the CERN council in late 1994. In 1996, the "technical proposals" of the ATLAS and CMS experiments were officially approved with an eventual project cost of 10 billion Swiss francs ($9.4 billion). In 1998, civil engineering work to excavate underground caverns to house the detectors for the experiments commenced. The original LHC Letter of Intent shows that the LHC was slated to be accomplished by 2002, but escalating problems with construction, budgets, and faulty parts all led to a long delay of seven years in its completion even as pressures on the LHC to get started kept mounting. Set to probe the structure of matter at the highest-ever energies achieved in the world, the actual operation date was pushed back every year. Finally, on November 20, 2009, two counter-rotating proton beams were successfully circulated in the LHC, with the first proton–proton collisions being recorded three days later at an injection energy of 450 gigaelectron volts (GeV) per beam. A few days later, the LHC became the world's highest-energy particle accelerator when protons in each beam reached an energy of 1.18 teraelectron volts (TeV). This exceeded the previous world record of 0.98 TeV, which had been held since 2001 by the Tevatron collider at the Fermi National Accelerator Laboratory, United States. "There's a great sense of anticipation here at CERN and at particle physics labs around the globe, and for good reason—we're about to open up the biggest range of potential new discovery that particle physics has seen in over a decade," announced Rolf Heuer, then director-general of CERN.[21] That anticipation was fruitfully rewarded in the experimental discovery of the Higgs particle two years later. But to analyze the meaning of this discovery, we must grasp the

framework of assumptions in which posing and solving problems acquires urgency and depth.

According to a familiar conception in anthropology and
sociology, the forward thrust of science can be understood only
pragmatically and locally, not rationally. As Paul Rabinow puts
it, "Without money and facilities there is no natural science."[22]
Hence, in the broadly shared values of science, accountability is determined, wittingly or unwittingly, through the vector
of calculability and interests.[23] If we confine our attention to
the ramifications of financial priorities and institutional practices, we will find that science grows out of local and particular concerns. It is undoubtedly true that questions of research
funding, the number of citations to published papers, competition among rival hypotheses, and other mundane matters are
grounded in the organizational rationality of the scientific process. The fact must not be lost, however, that it is also in the
conduct of experiments that meaning and efficacy of theories,
models, collaborative ventures, instrumentation, and R&D for
future projects resides. And if we trouble ourselves to listen to
the scientist, then it is in this anticipation of experimental discoveries that the self-conscious desire for novelty first emerges.
No doubt, money and facilities are necessary for science to take
place, but these are enabling conditions that have little bearing
on the content of science.

We have only to consider this in the light of our earlier observations on Standard Model physics. If the Higgs boson were
not detected or if every experiment drearily vindicates Standard
Model predictions, it is independent of how much power the
research community wields or the amount of money it has at its
disposal. I make this blunt assertion to point out that scientific
growth, whether on the plane of theory or experiment, discloses
quite remarkably that the most fundamental unit in science is

a problem or an issue. Rightly viewed, all episodic activity in experimental physics is bound together and made relevant in the problem: the setting up of experiments, the founding of collaborations, the building of colliders, and the conducting of future R&D, and these relations are not of succession but of simultaneity. The day problems cease, death ensues. The anxiety that informants express—that is, if the Standard Model Higgs is found, the field would be dead—recalls the tedium of habit and routine: in the wake of old problems, new ones must take their place.

This recognition, with its methodological ramification, rejoins the previous argument on the spirit of innovativeness that rules physics. Stories of major breakthroughs and exemplary laureates are shared and passed on with enthusiasm throughout the community. I found that these forms of storytelling, in many ways, touch on outstanding problems as links in a chain binding individual actors and the general destiny of physics. For example, Enrico Fermi's contribution to statistical mechanics after Wolfgang Pauli developed his exclusion principle or Hermann Weyl's analysis of discrete symmetries which led Paul Dirac to predict the existence of antimatter, are often presented as scientific leaps. As Dominique Pestre notes, "These stories are crucial to maintaining the values of the institution of science—the specificity and unique character of the knowledge it produces, for example, or the pivotal part played in its elaboration by the scientific method. They are essential to the smooth functioning and perpetuation of scientific communities."[24] Some of the stories shared also disclose the snares of politicking and self-aggrandizement. For instance, Carlo Rubbia's pursuit of a second Nobel Prize at CERN, which was close on the heels of the first one and at times bordered on unethical or opportunistic practices, is widely circulated in the community.[25] Or David Politzer's shrewd comments on the "Yang-Mills beta function," a calculation with only two possible signs, minus and plus, which nonetheless had three contenders to the Nobel Prize in Physics, is a remarkable public acknowledgment of the aggression and deception that physics seems to breed.[26]

These considerations, together with the necessity of "problem solving" that Thomas Kuhn has immortalized in his work as the everyday aspect of scientific process, make it imperative to recognize that conceptual issues form the decisive thread in science. With this as a point of departure, our appreciation for material practices or power relations is not diminished. On the contrary, conceptual issues are exactly where science breaks into practice and gives into the concrete demands of the day, such as which speakers should be invited to conferences, what topics will solicit future grants, and how to shore up citations of preprints or published papers. Conceptual issues bring an upswing in all domains of so-called material productivity. At the same time, the desire for innovations is not conjectural, but rather it is inwardly prompted, which also explains the disappointment of finding nothing but the Standard Model Higgs boson at the LHC.

PRESUPPOSITIONS

If we are right in thinking that experimental science proceeds by the constant transformation of its conceptual content, this implies that we need to unpack how it contains the criteria to validate new knowledge. From every angle, the interpretation of new discoveries borders on an enigma—that is, the enigma of deciphering the meaning of novel entities within the confines of existing thought. In this case, we need a sociologically dynamic ontology that explains how scientists make the leap in recognizing what is not known. Recall how Gaston Bachelard had urged us to understand that experimental science constitutes "a dialectical unity of reason and reality."[27] That is to say, there is a discernible rational impulse to reflect on the laws of the physical world. But there is also the predicament of technological realism, or the necessity of subjecting every theoretical reflection to experimental findings. Bachelard recognizes that the growth of experimental science is not chaotic and arbitrary. We may have reasons to smile at this archaic formulation, but we must not fail to inquire into the presuppositions that delineate the ground of experimentation. I insist on this point because the tendency that dominates currently is to portray with excessive fidelity the reality of science—that is, apprehending science in its immediate empirical existence. Questions about how experiments are organized around international collaborations or what political forces mediate between industry and science are rousingly posed, while a critical discussion of the conceptual elements that inaugurate the production of new knowledge are often passed by.

Ian Hacking has decisively shown the limitations of empirical reality in natural science by arguing that an experiment is a purposive intervention in the world. In his words, the primary aim of science is not only to understand the world ("representation") but also to change it ("intervention").[28] Accordingly, theories or models may approach the physical world in a conceptual manner, but experimental facts alone have the power to validate them. At the same time, he writes, experiments can generate unthought-of objects to spring into view—with hints of boundless plasticity of technology and human agency—such as electrons and quarks. The result is that as science expands, reality changes radically and qualitatively. Therefore, any scholarly analysis must seize the thought and action components of scientific existence (as opposed to aiming for a description of either one or the other). However, Hacking abstains from offering any suggestions on the method that would link the two tendencies of representing and intervening. In fact, he envisages experimental science as a "motley" of activities working precariously and almost independently without much relation to one another. For him, the development of the sciences does not have a grandiose Hegelian plan.

Rightly or wrongly, as soon as we understand the way the LHC is set up, some regularity or order begs to be recognized. I would say that the certitude of experimental knowledge has its basis in a twofold logical relation: (1) the total heterogeneity and separation of human nature and nonhuman nature, in which nonhuman nature can make its presence felt only when the scientist suppresses his or her own nature, and (2) the total heterogeneity and separation of theory and practice, where, this time, the modesty of the scientist is counterbalanced by the arrogance of the engineer, who can manipulate nonhuman nature, through will, for any purpose or end. This set of double determinations lays before us the exhaustive possibility that human will may readily be admitted in scientific *practice*, in which case the agency of the *subject* is maximal, whereas in scientific *thought*, it is the unqualified presence of the *object* that takes over. The logic of these oppositely wound impulses is not lost on LHC physicists. Giudice writes that in these LHC experiments, "one creates special situations, under controlled and reproducible conditions, to obtain quantitative information on nature's behavior."[29] That is, an experiment distinguishes itself in the immediate sense of listening to the voice of nature by *suppressing* the voice of the scientist. But then physics is also indebted for its productivity to "the human factor," for which Giudice looks to the "collaborations of about 2500 physicists [that] stand behind each of the two main detectors, and a slightly smaller number behind the other experiments put together. These people are ultimately responsible for the design, construction, and testing of every single component of these prodigious instruments."[30] These quotes throw light on the fact that the realm of instrumentation proceeds by *intensifying* human action, or the work of the engineer.

That technology is the pure play of human agency, which makes nature plastic, is well documented. From our genes to our gender, technology can produce anything. But what is striking is that experimental physics lives and progresses in this contradictory regime that separates human agency and scientific objectivity under distinct logbooks of pure and applied, or theory and practice. Predicated on this double splinter of relations, high-luminosity collisions take place in the collider and leave indelible traces, which the detectors, duly prepared, can capture and are subsequently read off by the physicists. That such vibrant traces should exist and be within the experimental community's interpretive grasp is entirely consonant with reason, but only after nature has made its mark, which leads to the reverence for such data and the affirmation of methods, such as observation, quantification, and measurement. Accordingly, the component of human action, when it is admitted as a presence in the technological support to the experiments, like getting the collider ready, installing magnets, or servicing high-energy beams, is neither neglected nor considered insignificant. Nothing is farther from physics than the desire to be primitive.[31] Human agency and technology are part of its form to solicit and conquer nature. Without the performance of the accelerator—its ability to achieve high-energy and high-luminosity collisions—experimental physics may never have disclosed its oracles on matter and energy. It is only with the human observer in abeyance, that nature runs boundless, provided the instruments have been assembled, techniques have been mastered, and materials have been ransacked. And when allowance has been made for fraud or illicit human intervention,[32] what remains is the half-innocent claim of matter presiding over the mind and the half-arrogant resolve of the human will crafting the universe. In no sphere of knowledge is this contrast clearer between physical nature and human control; in none perhaps has there been so much of both.

"Scientific practice is the only place where the object/subject distinction does not work," Bruno Latour writes.[33] In my view, this candid assertion should be taken as an admission of impotence rather than of pride. While it aims to acknowledge the failing of science in one vital quarter, and supplies the thrust for a critique, it unwittingly indicates the weakness of that critique and constitutes its chief difficulty. Properly speaking, Latour's critique hastens in the direction of manipulation of objects in scientific practice and, from that point on, he fails to inquire into the conditions under which nature is capable of a mathematical explanation as well as he addresses how it is subject to instrumental manipulation.[34] This creates the impression as if sociological accounts have no stomach for mathematical calculations or scientific theories, which are in fact no less human for being more austere. It is quite obvious that Latour's emphasis on the practical coupling of human and nonhuman is intended to push further the hybridity of reason and action.[35] But it is not destined to succeed for the simple reason that it proceeds from a self-deception about the very knot by which contrary forces of thought and practice are set in motion, and it settles for a tactical critique of mundane scientific practices—that is, detailing the labor of lab technicians or tracing the material artifacts produced during experimentation, isolated from theoretical impulses. A sustained critique of science necessarily involves parsing heterogenous logics. In short, if our ambitions are linked to tactical questions, we are thankful that Latour mentions in good faith the canonical form in which the operations of science take place. Most commentators have wisely pretended that the intersection of subject and object with theory and practice does not exist. If, however, we are keen to raise methodological questions in the space of the modern laboratory, we should recognize that human presence in the realm of practice by itself is no indication of science's worth or worthlessness. Fact and value or theory and practice are as far apart as possible, while means and ends are closely aligned. The consequence is a contradictory doubling where the human subject is soundly acknowledged in the realm of technology or instrumentation, whereas it must be omitted in scientific thought. While the objectivity of nature is emphatically affirmed in experimental thought, in the realm of instrumentation and technology, it can be effectively subdued and manipulated. On this grid of conceptual relations or presuppositions, particle physics reposes.

The importance of conceptual presuppositions was brought home to me from the moment I arrived at CERN. In the very first picnic, organized by the Theory Division in October 2007, a group of postdoctoral research fellows and visiting and staff physicists were discussing with much energy and verve the conflicting tendencies of mathematics and physics. Miguel Angel Vázquez-Mozo, a theoretical physicist visiting CERN from the University of Salamanca, Spain, mused aloud over a popular paper written by the acclaimed physicist and Nobel laureate, Eugene Wigner, called "The Unreasonable Effectiveness of Mathematics in the Natural Sciences."[36] In this paper, Wigner tells us that physics is an exact science, pertaining to a mind-independent external reality. Exactitude in physics is achieved through the accuracy and precision of mathematics. Pure mathematics, however, is a feat of abstraction of the human mind. (We see two trees or five chairs, but we do not actually see the numbers two or five.) Thus, it is a considerable mystery: How does physical reality respond so adequately to the language of mathematics, which is nonphysical or mental? Reflecting on this paper, John Ellis immediately brought the conversation around to complex numbers, or expressions involving the square root of negative numbers, which are even more perplexing to the physical senses than real numbers, such as two or five. He stressed that all students of physics know the part played by complex numbers. Yet he asked, can there be a bigger mystery that numbers for which there are no physical analog correlate so effectively with concrete, physical phenomena? Seconding Ellis's view, a postdoctoral fellow ventured an example, which I did not follow at the time, of how the mathematical representation of the unitarity triangle in complex plane constitutes the precise measure of the amount of CP (charge conjugation and parity) violation found in nature.

Participating in this conversation, I did sense that the relationship of mathematics and physics forms a serious puzzle to the informants. Undoubtedly, the origins of the puzzle hearken to the logical opposition of the mental and the physical. But it is also clear that the opposition does not lie outside of science but is intrinsic to it. As Wigner writes, "the enormous usefulness of mathematics in the natural sciences is something bordering on the mysterious and there is no rational explanation for it."[37] Einstein's observation is shrewder: "In my opinion the answer to this question is, briefly, this: As far as the laws of mathematics refer to reality, they are not certain; and as far as they are certain, they do not refer to reality."[38] Einstein's appraisal was echoed in a more tongue-in-cheek manner by Lyndon (Lyn) Evans, the famed project leader of the LHC, in a talk given at CERN's Main Auditorium that the formula for luminosity performance of the LHC contains pi in its denominator, a number with infinite decimal expansion, which makes him endlessly anxious about the alleged accuracy of the accelerator.

The reliability of mathematics is repeatedly invoked as the justification, legitimation, and sanction behind the success of physics. Yet it is always understood as a mystery, as a puzzle.[39] And this puzzle is not a limitation. Rather it constitutes the point of departure for all creative activity in particle physics. In fact, to resolve it would mean the dissolution of the discipline; the dualism of the mind and the world must be irrevocable for physics, as we know it, to take shape. This is the part that is expressed by the physicist's feeling of wonderment that "the miracle of the appropriateness of the language of mathematics for the formulation of the laws of physics is a wonderful gift."[40] Here Wittgenstein's aphorism comes to mind, "What you are regarding as a gift is a problem for you to solve."[41] The problem concerns the assumption of heterogeneity and separation of spheres. If the human mind and the physical world are habitually *presupposed* to be distinct and separate, then anything that suggests their interlocution would seem mysterious. Yet from a comparative anthropological perspective, one could argue the Achuar of Amazonia or the Kabyle of Algeria are not perplexed at the intimacies of the mental and the physical world.[42] From the first discussion with interlocutors at the Theory Division picnic, I gathered that the laws of nature may lack in consistency; numerical values of fundamental constants may vary over time; and models and theories may arise or dissolve unpredictably. Physics, however, is fairly certain about the demarcation of physical reality from mental activity even as it cross-fertilizes with human agency, in the realm of technology, to produce novel artifacts. This regime of theoretical speculations and experimental validations appears to distinguish modern science from, say, poetry or religion.

SOCIETY INSIDE SCIENCE

I say "appears" because an assessment of what distinguishes experimental science cannot be made without grappling with the debate on internal versus external factors influencing it. Here we encounter a paradox: to consider laboratory practice from within means taking cognizance of theories and hypotheses, computer simulations, and engineering parameters, which tell us why certain problems gain acceptance, while many others are discarded. Yet to understand the internal dynamics of a laboratory, we have to situate its endeavors in the wider context in which access to funding, promotion of career interests, or vying for international prestige become salient. Thus, we are faced with two conceptions of society: a society composed of a cultural identity, a sense of solidarity, or a shared language, as a unit *extraneous* to science; and, in contrast, a circumscribed domain of society coterminous with the research community, made distinctive by its professional beliefs, knowledges, and practices, which is found *inside* science. It is between these two conceptions that one finds bitter disputes fought over the external versus internal determination in much of sociology of science.[43]

Perhaps as a backlash to the controversy, the view that it is impossible to demarcate between strictly scientific and social determinations has gained in influence. Sharon Traweek's comprehensive inquiry into the constraints operating in experimental physics registers the point that lab budgets, research equipment, division of labor, and career trajectories correspond to the dominant cultural values of a nation. In Traweek's formulation, the substance of scientific research, as a sum of possibilities, can rarely be disconnected from external forces because the physics community grows out of the "national culture" of a people.[44] A similar view is articulated by Peter Galison that symmetry principles and computer simulations are inextricably "intercalated," and it is pointless to separate the inner dynamics of a research endeavor from the wider politico-economic environment in which it is situated.[45] The literature in this area is vast, and one could adduce several studies. The point is that since the clarion call to open the black box of science by questioning its technical content, there should be no going back on it.[46] The thesis outlined in this book is that the notion of society cannot be applied extraneously to science. It is more to the point to say that when we examine how concepts and issues are structured by the research community, we see society incarnate in a living, vital way. This accentuates another vital observation that a dissenting, critical spirit pervades throughout the community and reigns over it. As Fabiola Gianotti, formerly the spokesperson of ATLAS collaboration and currently the director-general of CERN once put it, scientific work only fertilizes in the soil of "barbarous and inhuman criticism."

During fieldwork I was made aware of how much physics abounds with fallen heroes, forgotten heroes, and false heroes. The trumpeting of a truth claim was always made in spite of others; inversely, one's inadequacy or failure was perceived to be a result of others' malicious interventions, like witchcraft accusations in so-called primitive societies. In March 2008, the UTfit collaboration from the Tevatron experiments at Fermilab announced "a clear (and clean) signal of New Physics" from CP violation in Bs mixing mediated by the weak force. Their results were based on the combined data from the two major experiments, CDF and DØ. In combining the data from these two experiments, the UTfit collaboration had to resort to "some wizardry."[47] Yet in spite of the steep challenges, they confidently claimed evidence of moving beyond the predictions of the Standard Model with more than 3 sigma deviations.[48] In the weekly Theory Colloquium, on March 19, 2008, one of the members of the UTfit collaboration, Maurizio Pierini, gave a talk announcing the advent of New Physics. The talk invited intense scrutiny from most of the physicists attending it. Seated in the audience, Guido Altarelli, of the eponymous Altarelli-Parisi equation renown, at one point asked with pretended modesty, "Can you teach me in what way the plots are consistent?" This question proved to the audience, by subsequent explanation, how hyperbolic the speaker's claim was.

The next day, another member of the UTfit collaboration, Guillelmo Gomez-Ceballos from the Massachusetts Institute of Technology, addressed the physics community in the Main Auditorium. He too argued the case for New Physics based on the combined strength of data, dedicated triggers, and robust analysis. The talk given in the Main Auditorium drew even more skepticism than the previous day's Theory Colloquium.[49] Pointed questions were raised on the way the errors from "systematics," or variations in measurement owing to calibration in instrumentation, were being combined from the two differently organized experiments. The questions showed an overwhelming distrust of the data and the claim of New Physics: "How do you take into account the non-resonant background under the Φ peak?," "Don't the pT cuts that you have, affect the angular distribution?," "You have half the statistics and yet you get the same number of events, one quote on the background and the same errors as DØ. How is that possible?" This last question from Tatsuya Nakada, deputy spokesperson of LHC-b experiment, evinced a laugh from the entire audience even as the speaker struggled to give a cogent response. Without derision in the slightest, what I wish to describe is that in the most solemn undertakings of science, drama slips in, reason and criticism take thrust, and questions of method and evidence occupy the foreground. So enormous was the discrepancy between the analysis presented and the data accumulated that the claim of New Physics was swiftly discredited.[50]

It would be a truism to say that the core competence of physics is embodied in the language of measurements and magnitudes. That this language allows physicists to be in conversation with each other as well as with nature is a given. What distinguishes it, however, is the tremendous room it allows for debate, dissent, and criticism.[51] A scientific breakthrough would not be enlightening if it did not show where one's peers went wrong. However, reflecting on conflicts and disputes should not lead us to conclude that the collective element of experimental culture, in the sense of hundreds in collaborations jousting in proximity, captures the properly sociological dimension of science. Marx's acknowledgment that even the solitary scientist is eminently social is a point to remember. He writes, "When I am active scientifically—when I am engaged in activity which I can seldom perform in direct community with others—then I am *social*, because . . . not only is the material of my activity given to me as a social product (as is even the language in which the thinker is active)."[52] This is true of theoretical physicists who most often work reclusively—that is, either alone or in clusters of two or three, and they keep going back to problems in their conceptual simplicity. Michelangelo Mangano, a CERN staff theoretical physicist, was fond of claiming that "old problems are worked by new methods." He also thought that older models often found success because of the consistent perfection of their results rather than their originality. Quite obviously, even those theoretical physicists who imagine that their models have nothing to do with applications or are not influenced by experimental results conceive their activity as belonging to a diffuse but vibrant communal life, even if carried out alone. Once it is admitted that all problems have their source in communal thinking, the notions of community or society go beyond a physical sense to insinuate a social practice that mediates between history and the individual.

Theoretical physicists, as a rule, have a great deal of confidence in their mental and mathematical abilities. The Theory Colloquium at CERN, held every Wednesday afternoon, was a fireworks display. The custom of talking over "coffee and cake" after the seminar could also contain the echoes of attacking thrusts and rebuttals in full strength. The speaker and a member of the audience might still be having an extended discussion on the topic presented at the colloquium until dinner in the cafeteria. There can even be much shouting back and forth. No computation or claim is let out of sight without first ascertaining the validity of its proof. And it is an important feature of scientific work that criticism be meted out brusquely to ensure that findings bear out in a nontrivial way. One speaker was advised by the colloquium organizer to wear his "boxing gloves" before the presentation because the "free use of fists" is permitted. The picture is not only dramatic but also attractive because it favors severity, contrarian thinking, and nonconformity. The standing and influence of theoretical physicists among experimentalists and engineers is considerable. They are known as problem-solvers and creatures of thought absorbed by the familiar notion of finding the so-called key to the universe. There was indeed far too much talk of solving mysteries and too little achieved by way of findings. The general acceptance of the conviction that decay has set in the field and that it needs a strong revival was conspicuous. From the standpoint of experimental physics, the decay is unquestionable—but not from the point of view of epistemology. To find hypostatized abstractions in every aspect of material nature (i.e., to assign constant values to changing physical processes) or in the power to predict is enough to acquit theory on the charge of worthlessness. When theoretical physicists speak of their "love of the unseen" as a characteristic of their science, it recalls a rare quality of Platonism. We must not forget that the reverence for mathematics is common to every branch of physics be it instrumentation, experiment, or theory. Most physicists can be found scribbling long calculations on a chalkboard at any hour of the day.

But even this is not the whole story. We have another reason to study the question of internal dynamism of particle physics. It revolves around the tacit belief of the physics community that its work crystallizes out of its inquiries and a wider context is of secondary importance, *except* when the community seeks to disseminate its results. Then, it invites the outside public to learn about the "secrets of matter," which it terms as outreach, and solicits "the interaction between science and society" by "promoting links between science and industry, contributing to the training of today's teachers and tomorrow's scientists, making everyone aware of today's scientific challenges, encouraging young people's enthusiasm for science, and combining the pleasure of discovery with the sharing of knowledge."[53] In the eyes of the organization, outreach involves a unidirectional flow of knowledge, from the scientific laboratory to the lay public. When a finding of physics is to be communicated, such as the success of a hypothesis or the observation of new experimental result, the persuasive force of the research community erases the boundary between the laboratory and the broader institutional context in which it operates. Suffice it to say, society is viewed by most physicists as an entity that can be detached from science or conjoined depending on the exigencies of publicity and promotion. I shall return to this discussion at length in chapter 5.

The truth of the matter is that the research community at CERN has little interest in the wider world and opts to reply to external influences with its own resources. But this lack of interest also discloses that experimental physics is not a static system or a finished ideology. It does not need prophets to launch it or skeptics to maul it. As long as it has the intellectual import of a problem, it is secure in its confidence. Its own intellectual issues are what renders experimental physics prodigal, intense, and relentless. If, on occasion, the laboratory seeks dialogue with the arts or humanities, it is to provoke a recognition for the "cultural element," which simply means assigning posthumous merit to residual aspects of their work-life or "the after six P.M. part," as Albert de Roeck, a physicist from the CMS experiment, condescendingly reminded me. The technical aspects, however, which involve determining the mass of the Higgs boson or explaining neutrino oscillations, will forever be outside the arbitration of any society except its own.

In the next chapter, the meaning of the internal dynamism of contemporary particle physics will come out more clearly when we consider how the experimental procedures crystallize with the dualism of fact and value to prepare the way for the breakthrough discovery of the Higgs boson. Anthropologically speaking, the notion of dualisms may sound like an outdated structuralist formulation. But this should not make us blind to their boundless breadth and analytical utility in the constitution of the scientific mind. While knives could be wielded in conversations, say, on the merits of string theory or the success of particle production in hadron collisions, the shared belief in the total separation of fact and value, or subject and object, has a subterranean existence and a strange persistence. These are beliefs followed without being justified. It is impossible to exaggerate the drama of scientists who are so filled with the passion of dualisms in both the substantive content of their work and the perspectives they apply to it. Certainly during fieldwork, I found in the dualisms an immediate ground to examine the social character of science. The larger question as to whether or not the dualisms constitute a self-sustaining paradigm are left open for the moment.

2

NATURE AND SIGNATURE

The Lord whose oracle is at Delphi neither reveals nor conceals,
but simply announces by a sign.

—Heraclitus

César Gómez once remarked, "physicists are Platonist only in credo. In actuality, they hold an absolute trust in the reality of things." Gómez is a theoretical physicist who also holds a doctorate in philosophy, which leads him from time to time to adopt an outsider's perspective to issues in physics. His baffling comment led me to reflect on the conditions under which an experiment may disclose the reality of things. In this chapter, I examine this question by way of the commonly used term *signature* in experimental physics. Characteristic patterns of decay formed by particles subsequent to collisions, such as a Higgs boson decaying into two energetic photons, are termed *signatures* and constitute the chief unit of discovery in particle physics. I will later take up a detailed exposition of this preliminary definition. For now, my concern is to introduce the notion of a signature and examine how it anchors the indisputable reality of things in science. To begin with, however, is it not a little paradoxical to approach things through signs, or the material through the mental? In positivistic thought, nothing is more radical than the opposition of things to signs. Things are concrete, exemplified by a materiality—or to use Hegel's expression, immediacy—whereas signs are pure values, arbitrarily generated and differentially interpreted.[1]

The paradox of the physics signature is that it is at once a thing and a sign, or a fact and a value. The LHC sees roughly six hundred million proton-to-proton collisions per second. But the bare fact of a collision tells us nothing because it means nothing. It is the attribution of meaning—as a signal—to the fact of a collision that makes it decisive. In this regard, meaning is not an outward garb subsequently added to a bare fact. Rather, in the very identification of a collision (fact), an evaluation is suppressed that inflects it with purpose and significance (value). To regard a pair of photons as a conclusive signal of a Higgs particle, or high-energy muons as a signature of dark matter, constitutes a mode of recognition involving two elements—thing and sign—that are so completely taken up into each other that although they can be identified as different in reflection, they participate intrinsically together. Not like anything that science can describe, the signature of physics revels both in being itself (a thing) and something other than itself (a sign). This bipolar orientation shifts modern metaphysics into a completely new light.

Underlying this investigation is an attempt to raise the signature from the interstices of the dichotomy of thing and sign to an existence as an autonomous and subversive element in a semiotic triad. At a time when we are revisiting perplexing questions and envisioning new elements of a postconstructivist evaluation of science, we may confidently give form to a nondualistic mode of inquiry that regards the universe not from the standpoint of things or signs but from that of relations.[2] Subordinate to no other aim save mediation, the postulate of relations expresses a powerful intellectual orientation. As numerous contemporary works attest, the overwhelming need to find relations between humans and nonhumans has led to the creation of novel nondichotomous concepts, such as cyborgs, hybrids, or assemblages, in rethinking much of the work accomplished in the natural sciences.[3] In the dispensation of emerging new sciences, these inquiries are distinguished for dispelling the illusion of pure categories like the biological, the social, and the material.[4] Varied in their focus, these approaches share in common an emphasis on the emergence of objects and on how these take on fluid avatars and enter new configurations.[5]

This path takes me slightly away from these recent developments. For in this rapid survey, I should point out one force that decisively governs the trajectory of relations. The links connecting heterogeneous objects, events, or practices are principally observed to follow a logic of instrumentality and contingency. Paul Rabinow and Gaymon Bennett note, "The dominant mode of rationality and purpose guiding the life sciences today is instrumental."[6] Moving away from "the imaginary desire of historical narration for coherence, integrity, totality, and closure, . . . we are currently witnessing a lively debate concerning the contingent, contaminated, local and situated making of science."[7] Donna Haraway has suggested that the commitment "after the implosions of technoscience requires immersion in the work of materializing new tropes in an always contingent practice of grounding or worlding."[8] A critical evaluation of contemporary technoscience, however, remains incomplete if we grasp the element of correspondence of nature and culture, or human and nonhuman, only in contingent practices or instrumental action. It remains onesided because the strength of the relation derives not from within but rather must be sought from the outside—the context—of motivations, which makes it provisional or partial.[9] The questions most intriguing to me are found in Émile Durkheim's framework, namely, (1) how *concepts* are forms of symbolic classification, and (2) what gives them their *necessary* or compulsory character.[10] Durkheim clarifies that the importance of conceptual associations is "not to facilitate action, but to advance understanding. . . . The Australian does not divide the universe between the totems of his tribe with a view to regulating his conduct or even to justify his practice, it is because, the idea of the totem being cardinal for him, he is under a necessity to place everything that he knows in relation to it."[11] The evaluation of the physics signature is a foil for considering in a comprehensive way Durkheim's insistence on concepts as instruments of knowledge that recall the necessary unity of mind and nature. The demand for such a discussion involving a leading physical science becomes especially alluring when we learn that "the concept of signature disappears from Western science with the advent of the Enlightenment."[12] Reemerging in the most exact of all post-Enlightenment sciences, this remarkable concept provides, as I aim to show, the means and material by which experimental science gives expression to the secrets of the physical world, where it shows a universal splendor and simultaneously exhibits a concrete and fastidious logic when its source is disclosed in human thought. To get to both these aspects, the material culture of the laboratory forms the indispensable starting point for our discussion.

HUMAN AND NONHUMAN

Amid much fanfare and publicity, the first beam of protons went into circulation in the Large Hadron Collider (LHC) on

September 10, 2008, at 10:28 A.M. It was a singular beam sent at injection energy of 450 gigaelectron volts (GeV), steered around the full 27 kilometers of the accelerator. Although no collisions occurred at the time, the event of the first beam generated considerable excitement at CERN. Crowds of physicists stood glued to the monitors watching the first operational LHC beam go around the accelerator. A little later, at lunch, I ran into Michael Doser, an antimatter experimentalist and deputy director of the CERN Physics Division. He asked me rather sarcastically, "So how does the first step in the social construction of the Higgs appear?"

Doser's provocative comment was meant to indicate the spuriousness of the claim of the social construction of nature when faced with its tangible materiality. Like most particle physicists, he was aware of Andrew Pickering's book, or certainly of its title, *Constructing Quarks*.[13] Doser's remark that theories of physics, which may well be social constructions, are materially constraining (and Pickering's work hardly ignores materiality or experimental reality) is a reminder of the tension between the physical and the human sciences that continues to exert an uneasy pressure.[14] The tension becomes clear when we examine how the physical component, like that of beams or collisions, is connected to the human element, of computations or calculations, say, and how the two simultaneously remain distinct.

In the physical register, the instrument looms large. The LHC is the world's most powerful particle accelerator. At a record energy of fourteen trillion electron volts, two counterrotating proton beams are made to collide head on every twenty-five nanoseconds. At four specified points of collision, gigantic detectors record the "data," that is, the product ensuing from the collisions. The site of each detector forms a distinct experiment pursuing specific physics goals. The two large experiments, based on general-purpose detectors, are ATLAS and CMS, which are designed to

FIGURE 2.1 ATLAS is the largest-volume particle detector in the world, weighing more than seven thousand tons.

Source: Image by Maximilien Brice, CERN.

investigate the widest range of physics discoveries, whereas LHCb and ALICE are specialized experiments that delve closely into the areas of flavor physics and heavy ions, respectively (figure 2.1).

Two elements of utmost importance reside on the human side: trigger and analysis. To obtain maximally "interesting" or atypical physics interactions, special parameters are used to make a selection. This selection is called a "trigger," which groups events according to bias, for example, a muon trigger would select events from collisions containing muons. This element of selection is vitally important because most of the particle interactions are considered "junk," because physicists have viewed these millions of times in previous experiments. In this case, they are focusing on atypical interactions produced in the LHC, because it is geared to unprecedentedly high levels of energy and luminosity. After the selection of data, or the trigger, the complex electronics of readout channels integrate millions of segmented data into a coherent description called an "event" and transmit the data to the computing grid for physicists around the globe to process and analyze. Analysis, then, constitutes the final stage of the combination and reconstruction of key events, and one in which the prospects of a discovery become distinctly conspicuous. Thus, it is not surprising that the analysis of data, or physics analysis, forms the most distinguished stage in experimental physics.

I started fieldwork at CERN in standard participant-observer mode in the ATLAS control room, tagging the T/DAQ (Trigger and Data Acquisition) Group headed by Livio Mapelli and later by David Francis. Within a few months, well-meaning informants impressed on me that trigger and data acquisition was simply a preparatory stage. The "real action" would begin, they said, after the data started emerging and people who did analysis would be at the forefront of the game, which is where I should be if I wanted my research to capture this frontline excitement. After some deliberation, I followed their advice and shifted my attention from trigger and data acquisition to event reconstruction and analysis. It was during the time spent with teams in physics analysis in the ATLAS experiment that I first encountered the term *signatures*, which suggested a unique intellectual orientation in the world of matter. Before I explicate this uniqueness, a feature of instrumentation deserves our attention.

To maximize the probability of collisions, the countercirculating LHC proton beams are divided into so-called bunches. Each bunch contains 1.1×10^{11} protons, and each beam contains 2,808 bunches. Every twenty-five nanoseconds, the proton bunches cross, resulting in about six hundred million collisions per second. An undeniable materiality or facticity clearly underlies experimental physics. The assumption so often set forth by naive realism as self-evident—that materiality explains itself—is an exaggeration. In the consideration of every single material fact is a concealed a prerequisite for its existence. This prerequisite is the gradation of purpose or significance. How is beam dynamics affected by collimation or why are superconducting magnets placed in front of the electromagnetic calorimeter? The characterization of purpose and significance built into such questions indicates that a perspective has been imposed on matters of fact.

Gradation of purpose or relevance is merely a conditioning mechanism. That is, in the functioning of technology, facticity and significance come together—or go apart—in accordance with exigency. To clarify, I am not arguing that operations involving beams and collisions do not require engineering skill or administrative decisions, factors that clearly suggest human intervention. For sure, an engineer is needed to turn on the beams or to design the beam pipe. But when the beam is running, no reference to a human observer is needed. That is, once a technology is instituted, it *functions* independently of the scientist or the engineer, whose presence becomes extraneous and is required simply for maintenance, safety, and repairs.

In contrast, in the case of a signature, fact and significance are intrinsically bound together as a *unity at all stages and modes of operation*. The relation to the human subject cannot be dislodged at any stage without losing the whole concept. What Charles S. Peirce said for signs—they "address somebody"—holds true of the physics signature. The signal is real, and this is absolutely crucial, but not because it is materially present in a collision. No, it is real because the physicist can recognize or receive it. The signature of physics compels our attention to the human subject. As the element of human recognition gains in strength, it does not abolish the relevance of the material, but rather it makes use of it to forge a unity. This coalescence of the human and the material and its anchoring in the conception of a signature is what I turn to next.

HIGGS → γγ SIGNATURE

It is easier to illustrate than to define a signature in experimental physics. Consider the following statement: "A two-body mass peak in the region of hundred GeV and above is the most robust signature one can hope for."[15] The application of this statement finds its most conspicuous expression in the decay of a Higgs boson into two photons (γγ) of definite mass. As an analogy, imagine a ball, whose material composition is not known a priori, getting hit and subsequently smashing to pieces. Now *if* the ball were made of crystal glass weighing, say, a hundred pounds, then according to the principle of conservation of mass or momentum, we should expect to find in the debris two conspicuous glass pieces of roughly fifty pounds each. Once we detect these two distinct pieces in the debris, we deduce that the original object in the collision was most likely a crystal object with a mass of more than one hundred pounds. Likewise, a particle collision, if it succeeds in producing a Higgs boson in a certain mass range, say, between 100 and 120 GeV, it would most likely decay into two energetic photons of roughly 50 or 60 GeV each, because of the principle of conservation of mass.

This analogy is useful but must not be pressed to an extreme. In relativistic quantum physics, decays are processes in which particles (such as a Higgs particle) *spontaneously transform* into other particles (like two photons), instead of decomposing or dissolving into constituent particles. New particles are really produced under the effect of field interactions based on the principle of the conversion of energy into mass ($E = mc^2$). That is, when fast-moving or energetic particles collide with each other, some of their energy is converted into the creation of completely new particles. For the Higgs particle, physicists were most keen to focus their attention on two particular trajectories: (1) the decay into two Z particles that further decay to four leptons (Higgs→ZZ^*→$\ell\ell\ell\ell$)[16] and (2) the decay into two photons (Higgs→γγ). Both of these modes of decay, which were dubbed "golden channels" by the community, were expected to disclose the production of a Standard Model Higgs particle. By the time collisions start taking place in March 2010, however, the diphoton decay channel had emerged as the more promising of the two and provided the preeminent signature for the Higgs boson discovery made in the next two years.

It should be clear by now that the term *signature* is used to characterize the decay products by which physicists identify the source particles, a kind of reverse deduction in which inferences are drawn about unknown particle states from observed final states. The task of deduction that a signature demands, however, is not easy because the signature has to be extracted from the background. Background refers to identical and competing processes that often fake a signal process. A potential source of background, as in this particular illustration of the Higgs decaying into two photons, is the photons produced by *bremsstrahlung*—that is, the electromagnetic radiation given off by accelerated particles when forced to travel in a curved path, like a circular accelerator. The photons from bremsstrahlung form the "irreducible background" and create a massive problem for physicists.[17] The problem is to figure out, on a statistical basis, which photons are emerging as a consequence of bremsstrahlung (i.e., the irreducible background) and which ones are decaying from a possible Higgs source (i.e., the signal).

While speaking to the convener of the Higgs search on the ATLAS experiment, Andreas Hoecker, I asked him how he dealt with the high photon background in the analysis of the Higgs boson signature. He replied rather nonchalantly that he did not believe in the concept of an irreducible background. Hoecker is a highly regarded figure in physics analysis, but I soon heard a few friendly murmurs of criticism in the community of his approach to background. If he did not believe in the background, how could he discern a signal? As skeptics were quick to point, because hadrons are composite particles, on a hadron collider in particular, an experimentalist must have a complete understanding of the irreducible background to obtain a relevant signal. Collisions involving hadrons are therefore rather messy and generate a lot of debris, that is, background. Skeptics also suggested that because Hoecker had previously worked on the electron-positron collider, BaBar, at SLAC, in California, where the collisions were relatively clean, he was prone to ignoring background contributions.[18] At this stage it was obvious that (1) physicists bring a continuity among experiments in the form of expertise from previous experiments and laboratories; (2) a multiplicity of views flourishes on the extraction of signals in the culture of experimental physics; and (3) the work one does has a distinct identity even in collaborations involving thousands.

With some hesitation, I went back to Hoecker and put to him the same concern on the extraction of a Higgs signal from the heap of photons, the background, generated in the LHC. He confidently replied that he was aware of the "conservative view in the community," but argued that while statistically plotting the decays, if a clear peak starts emerging from the "invariant masses of the energetic photons," it forms the "signal that these are from a Higgs." In contrast, the photons from bremsstrahlung with differing masses, the irreducible background, would be all over the graph, falling in the "tail regions" of a quintessential Gaussian distribution. The peak formed by the invariant masses of the two isolated photons in the final state against the overall shape of the distribution would give Hoecker and his team a clear signature of the Higgs boson.

Although I listened to Hoecker's explanation attentively, I confess that it took me almost a year to comprehend, and not without the aid of other physicists at CERN. In fact, listening to his and others' expositions on the way they extracted signals from collision data—as they filled my field notebooks with rough, sketchy histograms and plots—I was struck by how materiality turns away from itself while seeking itself. The task of probing matter is foisted on a vast semiotic terrain that involves recognizing signals, tracing them to initial conditions, and measuring background. It follows that the case of a material discovery, like that of a Higgs particle, can be settled only after isolating the photons that emerge from random bremsstrahlung (the background) against those that issue from a genuine Higgs particle (the signal). This, however, is an act of discrimination or judgment, which presupposes a subject, or more appropriately, a community of subjects. Without the recognition of the scientist immanent to it, nothing answers to the notion of a signature.

In this regard, we face a decisive turning point in the appraisal of modern science. The signature of physics presents us with an unmistakable case of nondualism between human and nonhuman, and it places before us a new angle of inquiry into science. As is familiar, several studies have called our attention to the entanglement of material-semiotic intermediaries in experimentation, emergent "interactions" of nature and culture, or the use of analogies and metaphors in the sciences.[19] In my view, however, this distinguished catalogue of studies manifests a lacuna in that exigency or contingency provides the impetus to the play of associations and meaning of science that is divined in the extrinsic rhapsody of assemblages and fractures, which are understood in an unending logic of means and ends.[20]

As it stands, this account is inadequate. To criticize science from the standpoint of contingency or exigency alone is to leave it half unread, because the possibility remains of "necessary relations" being obtained between concept and object, and the conditions under which science creates "its special universe of efficacious principles" are also possible, which has received little clarification so far.[21] This is the point from which I wish to advance current understanding. I argue that the signature of physics betrays a *necessary* unity of conception. The relation of sign and thing is not extraneously forged through metaphors or analogies. A heterogeneous mix of elements is not lumped together under the rubric of a signature. Operating by means of conceptual interrelations, the signature synthesizes individual features under a form and that form is held together by a perspective. As I've illustrated, without a perspective that recognizes in tracks of photons a signature of a Higgs boson, we cannot grasp how a collision provides the starting point of a new order of things in the precise moment of its decay. This perspectivization of nature, which the signature embodies, closes the gap of fact and value in a single stroke.

Thus, we realize the significance of my insistence that the signature exemplifies a unity as opposed to an assemblage or a hybrid. Donna Haraway frequently tells us that science is given to divergent influences, partial viewpoints, and fragmentary connections that have no necessary or essential relation to the world. The ground of its activity is purely exigent or external. The signature repudiates this approach. Instead of confusing diverse realms, it introduces distinctness into a confused obscurity. Instead of soliciting connections from the flux of materiality, it favors the concomitance that subtends and awakens human interest, which comes not by dint of (external) activity but by (inner) judgment and form. To ask if events are bound by connections is a concern of mechanism (e.g., Descartes or Locke), whereas our interest is in the principle (e.g., Leibniz or Kant). The signature forms the inner reason by virtue of which a discovery is what it is, thus affording an opportunity to grasp the source of scientific activity. The value of such an instructive opportunity reposes on not only the inner structure of a signature but also the community and its thought, as well as on the wider experimental tradition, on which I shall now elaborate.

DIFFERENTIATION AND SINGULARITY

The chief difficulty in recognizing a signal event is the background. This difficulty consists of knowing how to isolate and evaluate the strength, or significance, of a signal for which no strict rule exists. No foolproof method "works under all circumstances that can be picked out from a textbook," as Malcolm John, from the LHCb experiment, explained to me. "It is mostly a judgment call" on what procedures to follow for what data set. The judgment, which establishes this significance, is not a sum or an aggregate of particular instances but rather is a form of thought at once logical and statistical, which works in a kind of a double movement: a movement from confused data to a suggested meaning and from the meaning back to observational facts to which the suggestion had originally directed attention. This movement explains why problems at any level—whether they are hardware issues of alignment and calibration, or software concerns of timing, trigger, and analysis—can seriously affect the assessment of a signature.

The physics community considers the observation of a signature to be valid if a statistical significance of "5-σ standard deviations" can be obtained. The 5-σ level is simply another way of saying that a model or a theory has a 0.00003 percent chance of being false. Claims and counterclaims establishing a 5-σ discovery level are rampant as they are debated or dissolved in talks and publications. At seminars and conferences, I heard over and over again questions like: "How well have you estimated the errors from systematics?" "Are you basing your evidence solely against Monte Carlo simulations?" "To what degree can you distinguish a Standard Model Higgs from look-alikes?" In addition to background, informants explained, they have to guard against the so-called noise generated by electronics, such as the amplification of small electrical signals, as well as against statistical errors from limited event samples, detector effects like limited acceptance, and the resolution of measurements. A clear chain of tasks, specification of procedures, and recognition of skills operates, but the ultimate extraction of a signature requires a delicate and drawn-out process, much like finding a needle in a haystack.

A lot can be written to describe in detail the methods and procedures employed by experimental physicists in extracting signatures from the bulk of the background.[22] My aim is ascetic—that is, to inquire into the conditions under which the significance (of a signature) is gauged. Jacques Derrida identifies a tension at the heart of the general concept of signature—how to reconcile difference with repetition.[23] In his essay "Signature Event Context," Derrida looks at signatures in the everyday context and ascribes "iterability" as the primary specification. He writes, "In order to function, that is, in order to be legible, a signature must have a repeatable, iterable, imitable form, it must be able to detach itself from the present and singular intention of its production."[24] But equally, the signature is a singularity (i.e., the bearer of a unique identity), which Derrida recognizes but fails to elaborate on. I wish to pursue this question: What is the locus of the signature's singularity?

In the case of the physics signature, to be sure, unless a signal is repeated enough times, it would be difficult to recognize it. It is never, properly speaking, one signal or one event that experimentalists invoke. It is the peak around which a cluster of events coheres that gives meaning to a signature in physics. This requirement of basing evidence on a collection of cases is so prominent that statistical inference is treated as one of the yardsticks of experimental confirmation.[25] Although it is true, as Derrida asserts, that without the support of iterability or repeatability a signature would be less credible, recurrence is not a *condition* for its occurrence. This plain assertion is meant to put before us the recognition that characterization of singularity arises from differentiation.[26] Every experimental physicist attests to the merit of the statistical repeatability of signatures or to the need for counting and collecting like cases. What these physicists do not utter in words but that constitutes, or consumes, the whole of their professional existence is evincing the signal from its contrast, that is, the background. Nowhere does the meaning of the factual emerge with as much clarity as in the distinction of signal and background. To return to the previous example of the Higgs, repetition does not explain which ones amid the bulk of photons form the signal of a Higgs particle. How a material fact gains truth value lies not in the aggregation of events but in their differentiation. As opposed to Derrida's emphasis, I argue that it is not sameness as much as differentiation that explains the singularity and versatility of a signature. To summarize, I have shown that the signature exemplifies an indissoluble unity, and its meaning derives from differentiation, not repetition. Next arises the bigger question: How does the signature come to indicate definite material states of affairs? It would be rewarding to consider the input of semiotics, especially because they proffer the advantage of no longer needing us to inquire separately into decay modes, trigger criteria, or statistical significance.[27] Instead, we may focus on the concept of signal and its probatory force to ascertain how science could be assimilated in the general classification of signs.

THE CLASSIFICATION OF SIGNS

In Peirce's classification, for instance, we find that the relationship between signs and objects may take one of the following three forms: (1) iconic (based on similarity and resemblance), such as portraits; (2) indexical (based on contiguity and context), such as weathercocks; or (3) symbolic (based on cultural conventions), such as human language.[28] Ferdinand de Saussure's schema of signs is dominated by the category of the linguistic sign. Consequently, this schema emphasizes the arbitrariness of cultural conventions as the principal feature of signs.[29] As is well known, the chief danger in making the link joining the signifier and the signified arbitrary is the exclusion of the referential object from consideration.[30] More recent efforts by Umberto Eco, Giovanni Manetti, and Tzvetan Todorov have brought the arbitrariness of Saussure to grief.[31] Building on classical sources, most notably, Stoic logic, the authors have proposed a conception of sign that relates to its designata not as an equivalence in the sense that "P is identical with Q" or $P \cong Q$, but rather as an implication, such as "if P, then Q" or $P \supset Q$. This move is felicitous because the relation of inference or implication (e.g., "if there is a scar, there must have been a wound") works well for the class of so-called natural signs, such as indices and symptoms.

Yet, in some sense, we are left with two modes of equivalence and implication that operate at different levels, and either one could be accepted without the other. In their one-sided emphasis, a fundamental problem makes an appearance: a dichotomy begins to appear between signs that are considered to be *natural and necessary*, the reason of signification being contained in their very notion, and signs that are considered to be *cultural and arbitrary*, which derive their meaning from social conventions. This dualism of natural impulse and cultural intent, or *phusei* and *thesei*, appears in a number of approaches, including classical sources, such as Saint Augustine's or early modern works like Port Royal grammar.[32]

The chief objections raised against mutually exclusive accounts of sign behavior are that the grounds of division are insufficient or that overlapping criteria are employed, with the consequence that as many typologies as signs arise,[33] all of which lead Eco to rue that "there is a radical fallacy in the project of drawing up a typology of signs."[34] Although these objections are undoubtedly correct in raising our awareness to the complexity of sign behavior and the need for multidimensional typologies, a seductive alternative might be to accept and go beneath the apparent simplicity of the dualism to communicate something that has the possibility of transcending it. This is what the signature of physics achieves. The signs that disclose the structure of matter are necessary to the extent that a prior intrinsic connection exists between the signifier and they signified, but they simultaneously require a language to interpret this connection. This twofold aspect, of the sign as an instance and as an image of material generation, embodies the key to understanding how signatures form the thread through which the composition of matter is glimpsed and on which the very possibility of an experimental science depends.

Fortunately, this rare semiotic possibility in which a sign shares a likeness to a thing and also coincides with it, has a striking precursor in the medieval doctrine of signatures, articulated with great vigor by the Paracelsians in sixteenth-century Rhineland. With an ingenious eye for noncausal explanations, in the treatise "Concerning the Signature of Natural Things," Paracelsus offers a theory of dynamically related occult sympathies, or signatures, cascading through nature, from minerals to stars, which god has imprinted for humankind to intercept for their own benefit.[35] Every signature forms, he argues, an (1) outward vehicle for inner forces and faculties; (2) makes visible what is otherwise invisible and obscure; (3) depends "not only upon the principles of similarity, homogeneity, resemblance, correspondence and sympathy, but also equally upon the contrary principles of difference, separateness, heterogeneity and antipathy;"[36] and (4) "ultimately coincides with the created thing itself, insofar as it is understood as part of the divine plan and purpose of creation."[37] His central argument, made in the context of medicine, is that the truth of a disease can be confirmed by inspecting the symptoms, like rashes, boils, or fevers, that express them, without appealing to anything external. Likewise, the fact that at the root all of nature forms a unity is exactly what makes signatures such powerful operators. Indeed, in the concept of signature, Paracelsus never tires of urging us to not look for something behind the phenomenon, for it is itself the living manifestation of an order— that is, it reflects all the facts and relations of the universe.

This vitalistic conception of *signatura rerum*, or all things bearing signatures, orients them to meaningfulness and efficacy and acquires prominence in the works of later Rhineland thinkers, such as Valentin Weigel and Jakob Böhme. For Weigel, the fundamental bond between humankind and god is the knowledge of nature, which cannot be received from the outside.

Böhme is even more emphatic in advocating that "without nature God is a mystery."[38] Notwithstanding these lofty pronouncements, the metaphysics of signatures gradually breaks loose from nature to involve itself exclusively with mystical relations between humankind and god, which achieves impeccable clarity in the Tridentine dogma of the Eucharist in which the flesh and blood of Jesus are held to be "truly, really, and substantially" present under the appearances of bread and wine. Diverse contentions exist over the issue of the "real presence" of Christ, but what we nebulously grasp is the ontological possibility of a signature as something that is both itself (thing) and something other than itself (sign).[39] This summary of the signature's role in alchemy, cosmology, and theology reminds us of both its antiquity and continuity. Yet if the signature of physics deserves our attention today, it is not for being mystical, but for once again being the semiotic framework that makes nature expressive and experiment the hunt for these expressive effects.

Two objections to this account might be raised. First, the idea of conjunction of meaning and reference has been developed in the notion of indexicality and, therefore, the signature discloses nothing new. The objection is just, and indeed appearances favor the signature manifesting a contextual, or part-to-whole, relationship. I insist, however, that to understand a signature in the sense of physical proximity, like the case of indexicality, is insufficient. To return to the diphoton Higgs signature, the context alone cannot explain which photons, amid the bulk of photons, form the potent signal of the Higgs boson; a material fact gains truth value not through proximity but through selection and differentiation, which denote a relation of judgment.

The second argument, one favored by Foucault and Agamben, is the recognition that signature is "a form of similitude," which signifies through analogical resemblance.[40] Furthermore, although "the similarity is metaphorical," it has real effects. For instance, the thistle plant with its prickly thorns as a cure is efficacious against sharp and acute pains.[41] The question then comes up: How is a resemblance or a representation able to produce an observable effect? Looking at the issue from the standpoint of experimental physics, I would say that what distinguishes the signature is not so much the power of representation (*metaphor*) as the power of development of one representation into another (*metamorphosis*). A group of theory fellows at CERN, who work on phenomenology, patiently explained to me that signatures of quantum physics make matter intelligible through myriad transformations, which affect the values of physical quantities of particles, such as their masses or their lifetimes, as a result of their interactions with various fields. These transformations depend on quantum uncertainty and require several techniques of computation, called radiative corrections. These computations disclose how the effect of a transformation communicates itself, and if that effect is within the framework of the Standard Model or indicative of new physics beyond the Standard Model. Either way, and regardless of its type, every signature is characterized by the force of progressive transformations, which form the basis for the correlation of sign and effect.

Note that the heuristic value of signatures stands out most sharply when we consider that these are quantum events of discovery. Signatures are annunciations of particle states yet to come. Any overview of scientific semiosis must emphasize two closely related issues, namely, the promise of discovery and the hazard of error. Traditional semiotics offers little by way of explanation, except viewing these possibilities as the perfect foil for pragmatics.[42] The clarification of how the sign prepares the way to those steps in which it purely signifies itself or signifies itself to be an illusion is the final theme of this chapter. I will also conclude the original aim with which the inquiry into signatures began—that is, discerning the methodological import of the signature in the substrate of relations.

ERROR, TRUTH, AND POSSIBILITY

In a short essay, Bruno Latour revisits the issue of materiality and its relevance for science and social science.[43] Much is interesting and polemical in his essay, but the chief argument is fairly straightforward: restore the rights of the object, recognize its ability to mobilize orders of existence, and reconfigure language and society. The result, he writes, is "objectivity," which is "not a special quality of the mind, an inner state of justice and fairness," but simply the obduracy of objects, "how they object to what is told about them."[44] Latour recognizes the significant contribution of language to the scaffolding of objects, but he decides to place the accent on "the thingness of the thing." Undeniably, things form the substance of any laboratory science. Hits (of particles), jets (of strong interactions), or tracks (of electromagnetic charges) traced on the detectors can force the most recalcitrant or Platonist physicist back to things. To that extent, I do not disagree with Latour on the pervasive presence of things in science. I strongly disagree with him concerning their epistemological status in science or their implications for social science.

From my research, I wish to underscore that to comprehend scientific objectivity, one needs to consider not only the presence (or absence) of objects but also the possibilities of thought. As a consequence, this central tenet yields that a doubt can be raised about the objectivity of a scientific fact rather than being raised as a doubt about the existence of the object in question.

For example, in 1999, the Collider Detector at Fermilab (CDF), one of the experimental collaborations on the Tevatron accelerator in Illinois, reported the claim of a discovery of "New Physics" from an event of di-photons with a large amount of missing transverse energy. An event was observed in the detector with two photons and a large amount of missing energy in the final state. According to the rules of statistical significance, the event qualified as evidence for a discovery. But, as explained by Michelangelo Mangano, "Why do we not consider it as evidence of new physics? Because consensus built up in the community that . . . the evidence is not so compelling."[45] For further clarification I quizzed Albert de Roeck, the deputy spokesperson of the CMS experiment and one of my key informants at CERN. He replied, "Yes, there was a case of di-photons, but what did the event signal? Nothing. Of course, it didn't prevent theorists from going into a frenzy and proposing new models, but no attention was given [to those]."

De Roeck's dismissive response made clear to me that physicists' recognition does not simply convey an *interpretation* of a signal but forms part of the *conception* of a signal. If we take seriously Latour's proposal to demarcate between a "social sociology," dealing with the "symbolic," and a "physical sociology" that is attentive to "things,"[46] another question arises: On which side should the CDF signature of the di-photons with missing energy be placed? The real problem that remains unaddressed in Latour's essay is that if the existence of an object alone were decisive, then science would have no category of "error."[47] The category of error assumes particular relevance when we regard the process of formation of signal events: it holds a mirror to every object that presents itself as self-evident, and it shows once again that facts are meaningful by virtue of having a logical conception that, moreover, is bound up with the notion of truth.

During my fieldwork at CERN, I observed that discoveries and evidence were always debated against the expectation of truth. It was a familiar sight to see talks at CERN organized with titles like, "The Truth of the Top Quark" (February 14, 2008) or "How Charming Is the Truth: Search for Neutral Currents" (July 1, 2008). Even at the conversational level, truth featured prominently. After a seminar on "CP Violation in B mesons," most members of the audience came out of the hall heatedly discussing, "How can this be true?" When news came in from the Tevatron collider at Fermilab, in Illinois, on the exclusion of the Higgs mass in the window of 160 to 170 GeV, it created an outright commotion at CERN. A presentation given on the topic "Why I Never Believed the Tevatron Higgs Sensitivity Claims for Run2A and B" by Michael Dittmar on March 19, 2009, was particularly engaging. The discussion, following the talk, revealed the gamut of emotions that physicists are capable of—accusation (of fabricating the plots), defamation (of the Tevatron searches), and erosion (of trust in scientific collegiality). Above all, "Is it true?"

A cursory reading might suggest that in most informants' usage, the term *truth* coincides with reality. Entities, which are really out there, are true. But a sustained reflection, especially in relation to the work done on signatures, shatters any such simplistic notion of truth as a mere synonym or description of reality. For instance, originating in a common source, what permits physicists to distinguish photons decaying from a Higgs source (a signal) from photons of bremsstrahlung (an illusion)? In the daily work with which experimental physicists are engaged, sifting signals from backgrounds, the veneration given to truth forces the conclusion that truth is a yardstick, a criterion, to distinguish reality from illusion. No matter how routine and cliché, the sense in which truth means much more than reality is palpable. The manner in which physicists carry out their appraisal of signatures makes it plain that although they are obviously concerned with physical reality, they implicitly enact the vivid notion of truth as a measure of reality. In short, a reality is not given with which a certain knowledge is thereby gained, but it is the certainty of knowledge as true that constitutes the basis of their quest for physical reality. The more one focuses on signatures in laying bare this quest, the more obvious it becomes that the thought of this community never ceases at the checkpoint of reality. Instead, it gropes beyond reality and moves toward the possibility of thoughts and things.

At this point, we should recall how "missing transverse energy" (E_T) forms a generic signature that fuels the search for possibilities of physics beyond the Standard Model. Missing energy describes energy in the transverse plane that is not detected by a particle detector but that is expected because of the principles of the conservation of energy and momentum. Therefore, events with missing energy become generic signals of new physics processes, such as supersymmetry or extra dimensions. The upshot is momentous. Generally, we labor under the assumption that there are things, which are revealed as signs. In the case of unanticipated discoveries made through missing E_T, however, we have the opposite situation: a sign may stand for a thing that does not exist. Puzzling as this may seem at first glance, it neatly underscores that it is the power of possibilities that persuades scientists in the existence of objects not yet experienced that belies the myth that scientific objectivity is constituted in the transparency of "thingness."

Because of the overwhelming presence of signs in high-energy physics, Karin Knorr-Cetina describes its epistemic culture as moving "in the shadowland of mechanically, electrically and electronically produced negative images of the world—in a world of signs and often fictional reflections, of echoes, footprints, and the shimmering appearances of bygone events."[48] The decisive importance of her analysis is found in her recognition of the "sign processing machinery" as the distinguishing character of high-energy physics. Her shortcoming lies in consistently viewing signs as "fictions," "phantasms," or "chimeras," from which she deduces the world of high-energy physics to be a "ghostly self-enclosed system."[49] In my view, the question concerns not so much the ontological status of nature or facts, and to what extent they are to be viewed as constructed or described, but rather the perspective and the logic by which an experiment acquires knowledge of physical reality. Ultimately, if a signature of physics is stable, it is not because it is a thing, but because it is immersed in a determinate language from which it cannot be drawn apart.

I started my inquiry with the words of César Gómez on the trust scientists place on physical reality. Until now, however, I have stressed only the paradoxical character of the physics signature that defies any notion of brute physical reality. The sheer act of identifying a signature includes a change of form, an intellectual transposition that goes beyond physical reality. With this claim, I attempted to undermine Latour's argument on the power and autonomy of things. In contrast, signatures are not arbitrary assertions of signs or "fictions," as Knorr-Cetina contends, without an embedding in the order of things. By rejecting the position that physics reposes on the thingness of things, and its alternative that it is laden with arbitrary signs, we have reached a difficult position. We shall meet this difficulty by considering a third option: that of relations and their importance.

The argument is as follows: We have seen that the ground from which the sign shows the logic of the material world is not a single source but a Janus-faced relation of signal and background. Rooted in the same system, each side shows something specific about the physical world. Furthermore, what distinguishes them is the judgment that experimental physicists draw on in their evaluations. This judgment constitutes, as it were, the inner limits of the factual. But this does not suggest that discrimination of a signal is a matter of a few minds agreeing with one another. On the contrary, the act of judging marks the outer limits as the space of possibilities, with these configurations linking signatures to specific states of affairs. Because the set of possibilities is internal to the discourse, what it expresses is necessary. Finally, these two relational orders, the contextual, in which the signature is perceived, and the necessary, which engenders its genesis, form a crosscutting grid through which experimental physics caries out searches of new particles, like the Higgs boson.

Physics is so accustomed to separating facts from values that it has become incapable of appreciating even impersonal human judgment, except when propounding models and theories, for which personal credit is given, and in assembling the instrumental apparatus as an engineering feat, in which human presence is soundly acknowledged. Yet precisely at this juncture where discoveries in nature are claimed, we find the rather subversive concept of the signature, which enjoins on grounds of logical necessity (rather than, say, of ease of manipulation or functional coordination) a subject or, more appropriately, the community of subjects. However, from this investigation into signatures laden with meanings, given by a community and composed of a value, we must not conclude that science is personal or that physicists have a "feeling" for science.[50] On the contrary, what distinguishes the professional scientific subject from known feelings of subjectivity is the intellectual renunciation it exerts. Impartiality is not a negation of feeling, only an impersonal feeling. The signature of physics lays an insistent claim to the subject, who records a signal. And to make this record is to imprint a judgment.

3

ON ORIENTATION

I say, however: if you talk about essence, you are merely not-ing a convention. But here one would like to retort: there is no greater difference than that between a proposition about the depth of the essence and the one about—a mere convention. But what if I reply: to the depth that we see in the essence there corresponds the deep need for the convention.

—Wittgenstein, *Remarks on the Foundations of Mathematics*

Jorge Luis Borges tells the following story: Once upon a time, the King of Babylon orders the construction of an elaborate labyrinth. Taking pride in the bewildering complexity of the maze, he invites as his guest the King of Arabia to enter the labyrinth and explore its secret structure. The Arabian king is soon lost in the tangle of blind alleys. By dusk he cries out for divine counsel, with whose aid he ultimately finds his way out. Upon emerging, he compliments his host on his confounding construction, but slyly adds that he knows of a superior labyrinth in Arabia, which he would, Allah willing, someday allow the Babylonian monarch to behold. Several years later, after successfully waging war against the Babylonians, the Arabian king carries away the defeated Babylonian king to Arabia and takes him on a three-day long journey on camel-back deep into the desert. This vast desert—the Arabian king exclaims—is *his* labyrinth, unquestionably primitive but far more punishing. The Arabian king then unties the bonds of the captive royal prisoner and abandons him among the burning sands. Unable to find his way out of the desert, the Babylonian king soon perishes.[1]

Borges's parable depicts with depth and suggestiveness the tortures of space without differentiation. For space, Durkheim writes, "if purely and absolutely homogenous, would be of no use, and could not be grasped by the mind. . . . To dispose things spatially there must be a possibility of placing them differently, of putting some at the right, others at the left, these above, those below, at the north of or at the south of, east or west of, etc. etc. . . . that is to say that space could not be what it is if it were not, like time, divided and differentiated."[2] Durkheim reflects analytically, as Borges does allegorically, that uniform, continuous space is incomprehensible to consciousness. Navigation requires spatial distinctions, which lie behind the categories, top and bottom, front and back, inside and outside, or left and right. These categories define orientation. That is to say, categories of orientation are not reference points of some abstract geometry but are qualities of situation that arise from the precise awareness of our own body in relation to the immediate surroundings.[3] For example, we can think of "front" by which way our nose points, or to indicate a point downward, we may gesture toward our feet. Yet this contention is only half true and needs a further qualification. Even within these categories there are important differences at the outset. One can reasonably argue that the perception of our own bodies' front and back, or top and bottom, is easily made because these distinctions are noticeable. In the case of right and left, such as when we think of our hands, we do not find a significant visible difference. We may likely regard them as interchangeable. So, how then does one account for the distinction between right and left?

This deceptively simple question was thrust upon the French anthropologist Robert Hertz. It led him to exclaim, "What resemblance more perfect than that between our two hands! And yet what a striking inequality there is!"[4] The difference for Hertz is as mysterious as it is culturally compelling. To the right are reserved the attributes of purity, nobility, and strength, while to the left are assigned unfavorable connotations of evil, destruction, and even death. Furthermore, Hertz finds that the entire universe can be divided along the contrast of our hands, such as in so-called primitive cultures in which the right often represents the sky, stars, and the heavens; the left becomes the realm of femininity, hostile forces, the netherworld, and so on. Dualism is the concept that Hertz uses most often in characterizing the explanation that handedness receives in anthropological thought, with the suggestion that the moral and intellectual representations associated with right and left form a substrate encompassing "different and incompatible functions linked to contrary natures."[5] Note that Hertz does not seem particularly interested in the asymmetry of handedness found in natural phenomena, like on shells of snails or among amino acids, and the scientific attention these have received. Instead, he focuses on the symbolic and conventional character of right and left asymmetry with the idea of silencing naturalistic interpretations in which the slightly unequal skill and dexterity of the two hands is often explicated with anatomical reasoning, like the lateralization of brain development, which leads to inveterate muscle use. To grasp the preeminence given to the right hand, he argues, involves a careful scrutiny of how regularities are adopted as conventions, which is a dimension of usage and custom.

The remarkable importance of the conventional character to terms of *handedness* is brought home by Hertz through his analysis of language, in which expressions and meanings of the right-hand side, whether of the body or of the cardinal directions in space, convey a force that is in obvious contrast to the disquiet and aversion invoked by terms indicating the left. That said, we need to be careful about reducing the contrast of two hands to a rhetorical flourish or figure of speech. Hertz categorically writes that the hands are used "for the expression of ideas: they are primarily instruments with which man acts on the beings and things that surround him."[6] The relative stability of meanings and expressions are preconditions for the use of language. Hertz arouses my deepest reverence for the intellectual significance he assigns to a seemingly contingent division of right and left, while baptizing it as a "near-universal" form of classification.[7] Hertz shows how an arbitrary and cultural definition, like the designation of right and left, may at the same time meet the needs of a necessary and universal characterization.

This point is of profound and rare importance. With cultural institutions, we are constantly forced to appeal to context and contingency. It is a matter of common experience that jokes, mortuary rites, or moral injunctions can be explained only within a specific context. The criteria of humor or mourning depend on the circumstances. Conversely, the temptation to accept contingent conditioning in cultural explanations is so strong that changes in empirical circumstances can throw our concepts into confusion or rob them of application to reality. Universal definitions are generally not the instrument we use for anthropological constructs. Hertz, however, does not experience any such limitation. Instead, he boldly proclaims that the customary cultural difference of right and left pervades every sphere of human and cosmic activity. The embodied difference of right and left is the beginning of a series that stretches all the way to space and cosmos. How Hertz squares his account of handedness with the question of cosmology I attend to later, but it suffices to observe that his essay makes a double claim to be regarded as an epoch-breaking work: first, substantively, it forms the most eloquent expression of a theory of symbolic classification that stands at the crossroads of nature and culture; and second, historically, it forms the foundation for a conception that plays a significant role in structural analysis—that of binary oppositions. Yet, surprisingly, the concept of asymmetrical handedness, which enriches and transforms beyond itself an entire field of ritual and symbolism, subsequently vanishes from the anthropological vision.

Much of the work in this chapter is an effort to return to the polarity of right and left in particle physics to illuminate what it says about the interrelations of the human observer, social conventions, and the physical universe. The concept of handedness, or chirality, plays a leading role in the characterization of particles and their interactions. To be precise, the Standard Model of physics is referred to as a chiral theory, in which left-handed matter fields are treated differently from right-handed matter fields. Experimental observations have consistently found that beta decay under the electro-weak force produces particles that do not conserve the symmetry, or parity, of right and left. That is, in any such decay, more left-handed electrons are emitted than right-handed ones. This asymmetrical distribution is called a *parity violation*. It is also an interesting feature of the Standard Model that interactions involving gauge bosons of the weak force (the W+, W−, and Z bosons) with fermions also depend on their handedness (figure 3.1). That is, the $W^{\pm}$ boson couples only to left-handed fermions and right-handed antifermions. The Z boson couples to both left- and right-handed fermions,

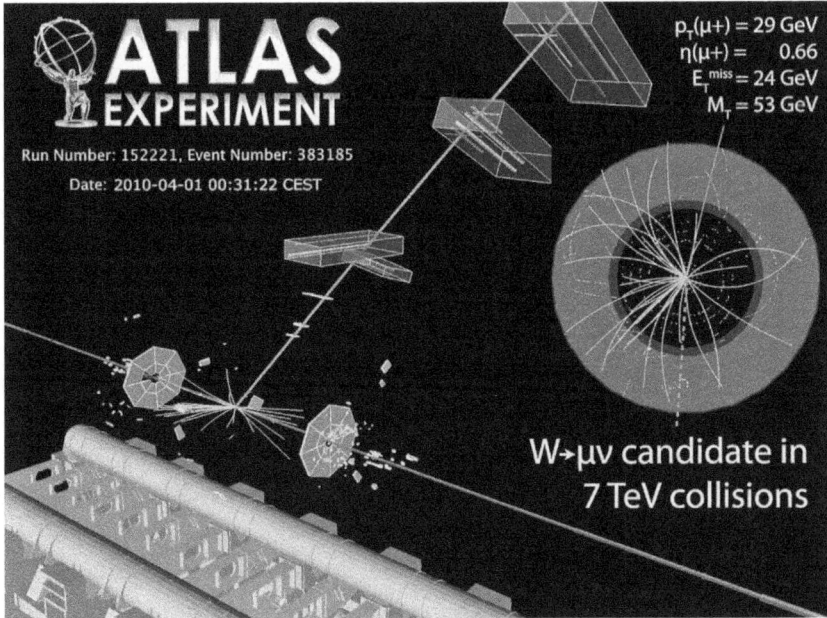

FIGURE 3.1 The extraction of a W-boson signature from the decay of two leptons in proton-to-proton collisions.

Source: Image by CERN, from the *ATLAS Experiment Blog*, http://pdg3.lbl.gov/atlasblog/.

but in each case, it does so with a slightly different coupling constant. The most striking illustration of right and left asymmetry in particle physics, however, is the case of neutrinos (and the antineutrinos): nearly massless neutrinos with an intrinsic spin of one-half, show a fixed left-handedness—that is, their angular momentum always points in the direction opposite to its motion, whereas antineutrinos always show up in the right-handed form. Performing a mirror transformation on a left-handed neutrino would yield nothing because no right-handed neutrinos can flip into the left-handed one. This is why neutrinos are said to be like vampires: they do not cast a mirror image.

The theoretical and experimental meaning of right and left asymmetry constitutes one of the most intriguing chapters in the annals of physics: why should nature rebel against the symmetry of right and left in elementary particles? Physicists find in this asymmetry an absurdity of nature. In the course of my discussions with them, I have seen the question take on answers ranging from conclusive to suggestive, which has the great merit of exposing the puzzle that handedness truly represents. This puzzle relates to how a subjective (or symbolic) designation of right and left shows up in subatomic physics as a complete specification of an objective (or material) condition. Reference has already been made to the danger of treating the distinction of right and left as comparable to top and bottom, front and back, inside and outside, and other similar categories of orientation, with which it is often concatenated. For unlike these distinctions, distinguishing right and left presupposes a human observer to endow space with a particular orientation.[8] As Hans Reichenbach pithily explains, the statement "Fifth Avenue is to the left of Fourth Avenue" is pointless and cannot be ascertained as true or false unless one specifies the direction from which one is looking at the streets.[9] So, let there be no mistake about it: for terms of handedness to make sense, the relation to the human subject can never be eliminated. This is what bewilders physicists.

Thus far, I have tried to bring to light concepts from within physics that express a *necessary unity* of opposed extremes. In the previous chapter, we paid attention to the classic dichotomy of fact and value and discussed how the signature of experimental physics shows their indissociable unity. In this chapter, I take up the concept of handedness for consideration to demonstrate how it expresses the necessary unity of the subject and the object. Yet before we begin, I stress that the intrinsic association with the human subject does not make the assignment of right and left a resolutely subjectivist matter. For although identical in shape, size, and structure, no matter how hard we try, we cannot wear a left-handed glove on our right hand.[10] Its materiality places some constraints on us. It is this double-sided nature to handedness, part subjective and part objective, that from a philosophical point of view has always struck those who considered it as intensely paradoxical, which sets it apart as a formidable problem in epistemology.[11] The problem turns on the question of whether knowledge of physical reality is at all possible without reference to a human observer.

The questioning of human presence in the "exact" sciences gives a distinct understanding of the latter. Almost all of physics proceeds by maintaining the strict separation and absolute heterogeneity of subject and object as a precondition to determine physical reality. For physics to possess any reasonable criteria of validity, it must be assumed that the object is heterogeneous from and constituted independently of the subject. Then, the task of analysis is to lay bare the form of physical object without any human mediation. Even quantum mechanics, which generated a significant controversy into the role of the observer, admits to only a *physical* presence of the body of the observer and does not include the *mind* of the observer as necessary to experimental outcomes.[12] As Werner Heisenberg observes, "certainly quantum theory does not contain genuine subjective features, it does not introduce the mind of the physicist as a part of the atomic event."[13] From my fieldwork experience, I can attest that faithfulness to facts "out there" constitutes the overwhelming feature of the physics community's thought and practice. During a light-hearted conversation with a few physicists in CERN's main cafeteria on "the scientific advantages" of tea over coffee, I complained that after a year of living amid physicists, my perceptions were gradually becoming biased. On hearing this, Thomas Mannel, a theoretical physicist visiting CERN at the time from Siegen in Germany, immediately protested, "But how can that be! We physicists are the most objective people on earth." Objectivity erupts from every pore of their belief system. Then, it may be reasonably asked, how does an objective science like particle physics involve itself with a perceived, felt, and embodied category like asymmetrical handedness? The ramifications of this question form the guiding thread of this chapter.

DISCERNING RIGHT AND LEFT

Properly speaking, what does a right-handed electron or left-handed neutrino imply in relativistic quantum physics? It first should be clarified that in quantum field theory, particles have their own fields. That is, particles are regarded as excited states, known as quanta, of a field. These fields exist everywhere in space and can be further subdivided into (1) matter fields, whose particles are termed "fermions," which include quarks, electrons, and protons; and (2) force fields, whose particles are called "bosons," which include photons and gluons. Fermions are the basic building blocks that make up all the observable matter in the universe. Bosons, in contrast, are responsible for the interaction of matter with physical forces, such as photons that act as force carriers of the electromagnetic force, or $W^{\pm}$ bosons that carry the weak force in nuclear decays.

Next, we come to the concept of spin. Spin is usually defined as internal angular momentum. Angular momentum is the magnitude used in physics to describe rotations, for example, the rotation of a ball or a dancing figure. If we were to choose a point on a rotating dancer's body, this point would define a circle around the imagined axis as it is rotating, and the angular momentum informs us how fast this part of the dancer is rotating. It is reasonable to argue, however, that the rotation of an object, like a ball, is not its intrinsic property. For instance, if we were to leave a ball on the ground, it ceases to spin. In relativistic quantum mechanics, however, spin is taken to be an intrinsic and indestructible property of all elementary particles akin to the property of electric charge or mass, which is revealed in the presence of a magnetic field. Moreover, this property of spin is generalized for particles that are conceived to be "point like," that is, something that does not take up any space or that has zero dimensions. Thus, material objects of modern physics, that is, the point-like particles lacking in spatial extension, are said to possess *intrinsic* spin. The idea of generalizing intrinsic angular momentum (defined for extended objects in classical physics) to unextended objects (the point-like particles in relativistic quantum mechanics) has been asserted by informants with all possible emphasis.

Finally, the combination of internal angular momentum with velocity provides the definition of handedness of elementary particles, under the twin notions of helicity and chirality. That is, every elementary particle has two possible ways to rotate, clockwise and counterclockwise. Because these are two physically dissimilar states, the angular momentum is a different vector in each case. Now consider the velocity of a particle. Here we mean the everyday concept of velocity, which can be represented by a vector indicating the direction of velocity. Combining the direction of velocity with spin in clockwise or counterclockwise direction yields a meaning of handedness under the associated notion of helicity, or the projection of a particle's spin in the direction of motion: When the direction of spin is aligned the same way as the direction of motion, then a particle is said to be "right-handed" or carrying positive helicity. When the directions of spin and motion are oppositely aligned, a particle is defined as "left-handed" or said to possess negative helicity. In less abstract words, a negative helicity particle rotates like a left-handed screw. According to an anecdote shared about the late physicist Maurice Goldhaber, in conference presentations, he typically would use the gesture of turning a door knob as if to open a door, thus rotating his hand in one direction, and repeating the gesture the other way to indicate the left-handedness of particles like neutrinos. Goldhaber's gesture, of course, draws attention to the fundamental fact that the assignment of handedness is made by applying the "right-hand rule." That is, when we curl our hand into a fist and extend our thumb upward, we find that spin is represented by the direction our fingers curl in, and our thumb gives the direction of orbital motion. The value of the right-hand rule is that it straightaway imparts a conventional flavor to the definition of handedness. That is, using the right hand to define the direction of alignment of spin and motion is a conventional and arbitrary assignment (in fact, the way in which any word stands for any entity is arbitrary or a matter of convention). That, however, is not the issue. What is being interrogated is would either of the handed objects, whether among particles, gloves, or shoes, make sense without an observer to endow them with a particular orientation?

The assignment of right and left to particle behavior depends on the particular definition of the direction of spin, what the community terms as clockwise and counterclockwise. The community could have just as easily selected the opposite convention and the resulting definitions of right and left would have been different. This relationship between direction and velocity conforms to the notion of helicity in which spin is simply displaced onto the notion of direction of velocity. Velocity, it might be recalled, is a relational concept that depends on the observer relative to whom we are measuring that velocity. In this case, handedness would vary depending on the direction from which we look at the particle and, therefore, it could not be a fixed property of particles. So, can a measure of handedness be fixed or determinate? The answer is yes. It is derived from Einstein's relativity principle. This principle renders the orientation of particles, in the form of chirality, a physically conserved or intrinsic property, just like charge, mass, or baryon number.

Let me explain. In special relativity, a maximum velocity exists—that is, the speed of light—and this velocity is the same for any observer in any frame of reference as it is tied to the very structure of space-time. It is taken to be a "fundamental constant" of space-time because no object or information can travel faster than the speed of light. Massless particles are, as far as kinematical behavior is concerned, equivalent to particles of light and are considered to be moving at the speed of light. The ramifying consequence is that the direction of motion of a massless object cannot be reversed under a mirror transformation. The speed of light is (1) the greatest possible speed and (2) independent of the reference frame. This is called the *Lorentz invariance*. Once it is known that the direction of velocity of massless particle cannot be reversed by a Lorentz transformation, an attempt is made to use this property to arrive at a definition of handedness that is intrinsic or absolute. Because a Lorentz transformation does not affect a massless particle's direction of motion, its helicity is absolute and is termed as chirality. In contrast, the direction of motion for objects with mass is reversed upon a mirror transformation. That is, massive objects are seen as moving in the opposite direction if we can match their velocity. For a massive particle, in principle, one could "Lorentz boost" (i.e., change of viewpoint) by a large velocity in the direction of a particle's motion, and thus reverse its direction of motion while leaving its direction of spin intact. Consequently, if we flip a massive particle's direction of motion, its helicity changes—that is, it would change from being right-handed to left-handed. This is what the physics literature means when it says that the "helicity eigenstate" of particles with mass involves both right- and left-handed components.[14]

To summarize, in relativistic quantum mechanics, handedness is defined with reference to spin and velocity, and it is distributed over the twin notions of helicity and chirality. Helicity is composed of spin (which is intrinsic) and velocity (which is relational). If a particle has mass, then the concept of right- or left-handed helicity is relational. If, on the contrary, a particle is massless, then left- or right-handed helicity is intrinsic, and this is denoted as chirality. The explanation for the distinction of helicity and chirality is related to the deep structure of relativistic space-time. Because massless particles are the only ones that move at the speed of light, and in special relativity, the speed of light is intrinsic and not relational, a real observer—who must always travel at less speed than the speed of light—cannot be in a reference frame from which the particle can reverse its relative direction. Therefore, immune to Lorentz transformations, the chirality of a massless particle is intrinsic to it in that all observers see the same handedness.

This discussion must seem prolix and dry, but it is also one that is adamantly set forth by physicists. What appears to be salient is that the application of spin to the direction of velocity gives the labels right and left an openly symbolic and conventional flavor. It is not possible to underestimate the methodological significance of shared beliefs over conventional definitions because it forms the starting point in the recognition of culture. Language is the most representative of this development. As highlighted by Saussure, linguistic signs are random sound patterns; only when they are stamped with social usage and conventions do they acquire meaning and purpose.[15] A similar position, in obverse, is expressed in Wittgenstein's declaration that a private language, that only one person speaks, cannot exist.[16] Meaningful communication is founded upon socially established common conventions. In extending these observations to the definition of left-right orientation in physics, we are inclined to ask on what foundations of nature is objectivity erected if we acknowledge that shared symbolic conventions decide the reference for predicates of handedness. We are concerned with the entanglement of nature and culture and with our difficulty to scrutinize them in the hard sciences.

Supposing the reader were to imagine communicating with faraway, alien inhabitants on the meanings of right and left if we did not share the same convention with them. As Richard Feynman seeks to argue, if we were communicating only by telephone (i.e., without drawing pictures or making gestures), we could describe our height or body size by appealing to some facts about the world that are the same everywhere, such as using the size of hydrogen atoms to express our height. But how would we describe to alien beings what right or left mean? To be sure, we could describe left as the side where the heart is. But what if their bodies are inverted? Or—it may well be argued—we could ask the alien beings to look at a polarized light beam rotating clockwise and specify clockwise rotation, or spiraling out motion, as left.[17] But that simply displaces right and left into other sets of distinctions in a series. Without agreeing beforehand on what is clockwise/counterclockwise, spiraling outside/inside, or positive/negative charges, we could not in any way explain ourselves to inhabitants of an alien universe.[18] The point is, as Feynman's argument demonstrates, terms of handedness are established purely by convention. This is the source of the problem of how to convey right and left unambiguously in the absence of a shared convention. Although we can imagine alien forms of life, and we can even imagine such alien creatures defining right and left for themselves, their definitions would be unintelligible to us in the absence of a common standard. Eccentric as it sounds, a common standard is available. It involves the formalism of electro-weak interactions and the experimental discovery of parity violation. The next section lays out this argument.

PREFERRED ORIENTATION

Before delving into parity violation, it is useful to go over the notion of symmetry. Symmetry is a property that describes how objects remain *invariant* even under certain categories of *transformation*, such as, rotations, reflections, and translations. When physicists exclaim that the laws of physics do not change if one moves to another room, they mean that the laws of nature are symmetric under a transformation in space.[19] The use of symmetry, however, has two levels. One level is the reckoning of symmetry in a geometrical or physical sense, as just described. To advance another example, rotating an equilateral triangle by 120 degrees around its center results in complete identity, or symmetry. We must distinguish a second sense in which symmetry is attributed to laws, rather than to material objects or phenomena. Some of the fundamental symmetries in physics are space-time symmetries, like the Lorentz transformation, where the *laws of physics* remain unchanged under boosts and rotations. Consequently, because symmetry expresses invariance under a change, it exerts a deep influence about why contemporary physics identifies symmetry as a robust measure of objectivity. For example, the statement "body A has velocity V" cannot be a law of nature because for observers in different frames, body A may show different velocities. The statement "the relation between force and acceleration of a body is its mass," can be a law of nature event though different observers may measure different forces and different accelerations, because their ratio would, in any reference frame, be equal to the mass. Thus, once a symmetry is identified, the physicist can deductively infer which properties of a physical system are invariant or conserved and therefore are objective.[20] The point to emphasize is that symmetry is densely interwoven with objectivity. In substance, it follows that laws of physics should be indifferent to a system and its mirror reflection. The idea that fundamental physical interactions would make it impossible to distinguish right from left or clockwise from counterclockwise is fostered by the term *parity*.

Parity is a concept that is related to (discrete) symmetry.[21] Parity refers to how subatomic particles behave if the spatial configuration is changed under a mirror reflection. Spatial inversion is one in which a physical object or process is described by a new set of coordinates that are negative signs of the original, like in a kind of mirror image where x goes to –x, y goes to –y, and z goes to –z. The direct consequence is that under such a spatial inversion, right-handed particles change into left-handed ones, and vice versa. Conservation of parity simply means that in replacing a right-handed coordinate system with a left-handed one, the results of a fundamental experiment would not change. If the laws of physics were to change when all the signs of spatial coordinates are flipped, it would be deemed as violation of parity. In classical physics, all the fundamental forces were *assumed* to be invariant under spatial inversion (i.e., as preserving parity). By the 1950s, however, a suspicion was raised as to whether it was conclusively known if particle interactions involving the "weak" force, the force responsible for radioactive decay, preserve parity.[22]

The situation was like this. Postwar physics had received a stimulus from the setting up of gigantic atom smashers in laboratories at Berkeley (the bevatron), Brookhaven (the cosmotron), Birmingham (the synchrotron), and CERN (the proton-synchrotron). These machines saw a plethora of subatomic particles being blasted out. The new particles also brought new problems. One such problem was the tau-theta puzzle (τ-θ) in which these two particles with identical masses and lifetimes were somehow observed to be decaying differently.[23] A number of experimental tests were conducted to measure the branching ratios, production cross section, and scattering properties, which would explain why the tau decayed into three pions while the theta decayed into two pions. This was to no avail. In 1953, theoretical physicist Richard Dalitz suggested that if conservation of parity held, the theta should have parity of +1 and the tau of –1, and the two could not be the same particle. The clear difference of parity would explain their difference in decay modes. To invoke a difference in parity for two particles identical in every other respect, however, seemed to be poor practice. The way out of this predicament was suggested by an innocuous question at a conference that entailed significant unintended consequences. In February 1956, at the Annual Rochester Conference in New York, Feynman, at the behest of Martin Block, raised an open question, namely, how can one be sure that weak interactions conserve parity? To voice one's misgivings regarding the conservation of parity was a novelty at the time because it was widely assumed that parity is conserved, that nature does not observe any distinction between right and left, and that any experiment would give the same results whether conducted in the actual physical world or in the kind of Alice's inverted, mirror world. Immediately after the conference, however, T. D. Lee, of Columbia University, and C. N. Yang from Brookhaven National Laboratory, got together to cast a critical eye over the specific question of parity in weak interactions.

On June 22, 1956, Lee and Yang submitted a paper to the *Physical Review* journal titled, "Is Parity Conserved in Weak Interactions?" The editor of the journal, Samuel Goudsmit, felt troubled by the audacious claims of the paper and protested against the use of a question mark in the title. Thus, the paper was published with the revised title, "Question of Parity Conservation in Weak Interactions."[24] In this paper, Lee and Yang argued that the law of parity conservation had been wholesomely tested and proven for the strong and electromagnetic interactions, but no experimental evidence was available to either confirm or refute parity invariance among the weak interactions. Weak interactions constitute the nub of radioactive decay. They proposed two different solutions: one, that particles exist as "parity doublets"; and two, and much more revolutionary than the first, that the long-held idea of parity conservation may be violated in particle interactions involving the weak force. They suggested a number of experimental tests, which would determine whether or not the weak interactions discriminated between right-handed and left-handed processes during radioactive decay. A year later, C. S. Wu assembled and executed a highly complex experiment involving polarized cobalt atoms set against a uniform magnetic field to test the orientation of particles emitted during radioactive decay. The idea was to examine the emission of decaying electrons with the magnetic field oriented in opposite directions.

One would have expected the electrons to be emitted uniformly in all directions because under the assumption of parity conservation, the emitted electrons from the decay of Cobalt[60] should show no preference for either left or right helicity. The experiment instead disclosed the startling fact that more electrons were emitted in the direction opposite to nuclear spin rather than in the same direction, that is, more left-handed electrons were emitted than right-handed ones. The asymmetrical emergence of particles preferentially in one direction rather than the other, termed *anisotropy*, in cobalt decay conclusively established handedness in the laws of nature.[25] I do not portray the experiment in greater detail because my aim is to underscore that experimental validation of parity violation forms the key point at which the division of right and left, which is conventionally defined through the right-hand rule of angular momentum, acquires the character of an objective fact.

But how strange that a concept should keep its symbolic, variable character and freeze into a materially discovered, objective fact. After all, we must bear in mind that the process of defining right or left chirality, for both massless and massive particles, cannot be done without consciously choosing a particular spatial orientation as a convention. Then, to witness nature converting an arbitrarily chosen human definition into a grounded, substantial, and absolute fact is astonishing. From the earliest conversations on violation of parity with physicists at CERN, I noticed the shock and awe they expressed at not only seeing the fall of a cherished space-time symmetry but also for pressing toward a larger question: How can impersonal laws of nature distinguish between left and right? The question is by no means academic. Quite apart from the experimental context in which theoretical assumptions are almost expected to be upended, the failure of right-left symmetry is about the conceptualization of qualitative categories in embodied consciousness, or how we orient ourselves in space in relation to our bodies. This embodied meaning of handedness links up with the larger question of orientation—that is, about the mind and how it discriminates—which most physicists instinctively grasp.

When T. D. Lee was visiting CERN in 2007, I asked him to describe his role in the historic discovery of parity violation and, in particular, what he thought of the asymmetry of handedness measured against the ideals of science. He replied emphatically: Whosoever wishes to philosophize about the form of explanation that parity violation invites, should do so without the error of attributing to it sensory impressions; we cannot escape from the fact that through the experimental discovery, nature has revealed her mind but, he added in an undertone, that it is a "universal mind." Lee proceeded to describe how the discovery of right and left asymmetry in weak interactions resulted from "the weight of facts they were faced with in nature." He concluded that these facts simply tell us that some particles decay through the weak force in unequal measures. Now, with this explanation in view, if we return to Feynman's puzzlement of how to communicate a definition of right and left unambiguously to an alien species on a distant planet, then it is obvious we have an answer. We simply ask the alien species to carry out experiments with neutral kaon decays. If the alien inhabitants manage to produce in their laboratory long-lived neutral kaons, then by observing the asymmetry of decay products they could clearly discern right from left. What previously seemed like a matter of pure difference based solely on a convention has now transformed into a radical asymmetry with support from the laws of nature.

We clearly touch a critical nerve of the problem. Parity violation is not about an observed asymmetry of material particles spinning in a specific direction, like our hearts being on the left when they could have easily been on the right, or the sun's movement through the galaxy in a helical path when it could have orbited in a helical path of opposite direction. The physicists insisted that these would merely amount to contingent facts. The point is more momentous. It is about the laws of nature showing a preferred orientation. As Edward Witten, one of the princi-pal architects of string theory and a rare physicist to be awarded a Fields Medal, expounded in an interview to me, "parity viola-tion involves the existence of a preferred orientation of laws in physical space." But to ascribe the expression of a preferred ori-entation changes the understanding of nature radically and qual-itatively.[26] Because spatial orientation of right and left is always relative to a point of view, it subverts the Nobel Prize–winning physicist Steven Weinberg's claim of "a chilling impersonality" to the universe.[27] In discriminating between right and left, the impersonal universe shows it has discernment. It is this problem that lies at the core of the puzzle that asymmetrical handedness represents: If physics presupposes the separation of mind from matter, or subject and object, then how can it base a physical universe with a preferred orientation? If it does not, then what sustains its claim to pure objectivity?

POLARITY AND UNITY

The problem of handedness has a long and distinguished gene-alogy in Western epistemology. It is Immanuel Kant who first ponders over the puzzle that the two hands pose as a chal-lenge to the question of space from which his whole "critical" philosophy was to evolve.[28] Our hands, which are so similar in shape and equal in size, have an inner difference: one hand can-not occupy the same portion of space as occupied by the other hand, although its reflection can. The difference, Kant argues, lies solely in how the hands are oriented in space. The next step in his argument—very important for its implications for embodied cognition—is that the ground on which orientation rests is the discerning subject. It is our awareness in qualitatively feeling the discrimination of directions, mediated through our body, that is the key to determining spatial orientation.[29] While scientific knowledge was increasingly reducing sensory distinctions to metrical properties, Kant recognized that the concept of orientation resists this reduction and reintroduced the necessity of considering form, or quality, along with magnitude, or quantity. He then made the more audacious claim that even a solitary hand in empty space would bear the definite identity of being a right or a left hand. This claim went against the grain of the then-prevalent thinking, which had been made the dominant view by Leibniz who had argued that space and time were not *things* in which bodies or events were located but rather were *relations* that could be obtained between actual, and even possible, objects and events. "Space is an order of co-existences, as time is an order of successions," he wrote.[30] In fact, Leibniz had conjectured that if everything in the world were shifted from east to west, retaining all of the relations between bodies, the universe would still remain the same.[31]

The puzzle that orientation categories pose for the concept of space was well known for three hundred years, but no one had converted this knowledge for particle physics in more meaningful ways than Hermann Weyl. Weyl picked up the threads of the debate from Leibniz and Kant to refute Kant's naïve assertion that even a lone hand possesses a determinate handedness. In a vein similar to Leibniz, Weyl argued that whether a hand is right or left cannot be determined by any of its own properties but only by considering it in relation to another hand and checking the respective orientations of the thumbs can it be decided whether it is left or right.[32] He went on to suggest that "scientific thinking" sides with Leibniz while "mythical thinking" subscribes to Kant's position "as is evinced by its usage of right and left as symbols for such polar opposites as good and evil."[33] Weyl's focus is unquestionably on the requirement that the association of a direction is always understood in a conventional sense.

The anthropological record shows that in mythical consciousness, space is often conceived as qualitatively differentiated with regions above and below, right and left, downstream and upstream, which acquire distinct meanings because each region or direction is imbued with specific values of the sacred.[34] Scientific articulation of space, in contrast, commences with the homogeneity and isotropy of space—that is, space is the same in all locations and in all directions. Giordano Bruno was one of the first to assert unequivocally the homogeneity of the universe.[35] René Descartes dealt the ultimate blow and proceeded toward a thoroughgoing geometrization of space, whose essence is extension, writing that "extension in length, breadth and depth is what constitutes the nature of corporeal substance."[36] From this, he deduced the fundamental tenet of his epistemology: matter is what we comprehend through extension. Leibniz vehemently denied the Cartesian postulate that extension alone make up the essence of matter. If matter were pure extension, Leibniz objected, then it would be incapable of grounding any dynamic features such as force, resistance, or activity.[37] Dissatisfied with Cartesian attempts to reduce space to an abstract geometry, Leibniz embarked on devising the science of topology, one that was based on situation (*analysis situs*) rather than on abstract metrics and magnitudes. In this lively exchange of views, we become aware of how the antithetical modes of causal-mechanistic and finalistic-teleological reasoning have jousted side by side for more than three hundred years. On the question of spatial orientation, philosophy has been in a state of flux.

We next return from these far-ranging generalizations to a specific methodological point. Since medieval scholasticism, intense debates have centered on the categories of substance and attribute, with near-total neglect of relations.[38] Empiricist approaches have tended to accord priority to substance over relations by devaluing relations as a creation of the mind and, therefore, as unreal or groundless. John Locke, for instance, recognized that knowledge of relations constitutes the largest field of our knowledge, but he denied them any significance in his epistemology. He viewed them as "extraneous and super-induced," an artificial product of the human mind.[39] Locke's whole theory sought the repudiation of mind in the conception of knowledge. Descartes had taken a similar but inverse route in the sense that he denied sensory experience any formative principle but, like Locke, ended up positing a thoroughly dualistic universe with mind and body as two distinct and mutually exclusive substances. The challenge, of course, is knowing how to explain the interlocution of an immaterial, unextended, indivisible mind (*res cogitans*) with a material, extended, and divisible body (*res extensa*). Within the framework of either rationalist or empiricist thought, the problem is unsolvable in principle.

This is not the historical problem that Descartes faced in 1644 or that Locke faced in 1690. I should like to revisit Hertz, who explicitly seizes upon the distinction between right and left as one of "the essential articles of our intellectual equipment," which overcomes the structural limits of "one or the other traditional doctrine concerning the origin of knowledge."[40] He declares that "those who believe in the innate capacity to differentiate," by which we may safely assume, he means rationalism, "have won their victory," since representations of right and left "are true categories, anterior to all individual experience." At the same time, he believes, "the empiricists were right too" because the categories of handedness are "subject to change and dependent on complex conditions."[41] In the balance, he discovers that the preponderance of right-left asymmetry could be understood only in concentrated form when it is revealed as an expression of a universal relation of differentiation. Hertz repeatedly emphasizes that binary differentiation, or dualism, alone can provide a synoptic view of the association of right and left with the oppositional values of pure and impure, light and dark, and male and female, which characterize entire cosmologies of primitive societies. He states, "the obligatory differentiation between the sides of the body is a particular case and a consequence of the dualism which is inherent in primitive thought."[42] If we set aside for a moment the aspect of primitive thought, Hertz's insistence that we accustom ourselves to the idea of binary differentiation is, at bottom, an acknowledgment of the logical coherence of the world. The general drift of his analysis is that the distribution of right and left distinction across cultures already insinuates an identity between concept and reality.

But this is just one aspect of the argument. We must not forget that what is decisive for Hertz is that the differentiation of right and left cannot exist as a "pure" abstraction involving only the intellectual sphere; it is a concept centered on our living, thinking, embodied selves. In brilliant and compelling words, Hertz poses, "How could man's body, the microcosm, escape the law of polarity which governs everything"?[43] A few lines later he responds, "If organic asymmetry had not existed, it would have had to be invented!"[44] In this climactic edict, he gives us the living relation, not just the logical relation, of the dialectic that captures the experience of polarity engendering from within itself a unity of mind and matter, which is fulsomely revealed in the form of cosmology. That is, because we think and act by means of distinctions, such as using the right hand to take an oath or the left hand for burying the dead, investing each with masculine and feminine virtues, or values of the sacred and profane, the breadth acquires a coherence that presses forward to

encompass the cosmic social order. Concrete analysis of handedness thus implies a relationship to the cosmos as a whole. In
Hertz's words, "society and the whole universe" participate in the
asymmetry of right and left.[45]

Hertz's greatness lies in the fact that he sees the anthropological significance of handedness more incisively than any
other scholar, and he recognizes the defects of those who think
we are dealing with a sporadic and isolated empirical phenomenon, thus, underestimating the widespread occurrence of
asymmetrical handedness. More interesting, perhaps, is how he
rejects a narrowly conceived lateralization of brain function as
a naturalistic explanation for the asymmetrical treatment of the
hands. His methodological emphasis on the necessity of differentiation, as disclosed in symbolic classifications, leads him to
elevate the humble concept of handedness to a quasi-universal
status, which permits cosmological comparisons between different cultures. That should not be surprising because it is
his view that things and beings exemplify not only empirical associations but also necessary connections. To clarify any
misunderstandings, I should remark that no direct, fixed path
leads from the views of Hertz to the situation found in particle physics today. Admittedly, in particle interactions involving the electro-weak force, no opposed sentiments or values are
attached with the right or left hand. It would be naïve to believe
that if vector bosons of the weak interaction couple to left-
handed quarks and leptons, it implies that nature inherently
prefers the left orientation. There is no question in physics of an
absolute preeminence of either a right-or left hand similar to
Hertz's formulation. It would also be senseless to suppose that
the attributes of spin and chirality are rooted in a binary scheme
of the kind exhibited by the polarity of sacred and profane that
Hertz outlines.

My analysis of handedness in physics draws on Hertz at three levels. First, I have already identified the philosophically significant side of this approach, its methodological aspect, namely, the fact of contrast—right and left hands—in the face of their obvious physical similarity, which immediately opens up the possibility of perspective and discrimination. We may generalize a little expansively: without relations of difference and otherness, even a concept of physics does not acquire form, at least, in the rare case that handedness represents. I repeat it is rare because concepts in physics refer to matter's own or self-subsisting properties; the characteristically perspectival or relational meaning of spatial orientation obfuscates that picture. Second, we raise the question of sociological unity, or how collective conventions inform the division of right and left. In particle physics, the process of assigning right or left helicity, whether for massless or massive particles, cannot be done without consciously choosing a particular orientation of space as a convention. This decision about what orientation to choose as the standard convention, the right-hand rule for angular momentum, forthright involves the community at large. Once it is recognized that the human subject and conventions of a community play an integral part in the formulation of chiral phenomena, it puts a question mark on the proclaimed objectivity of physics. Third, I have reflected on the experimental discovery of parity violation by virtue of which the laws of nature are said to treat fundamentally differently objects of opposite handedness. This extends the discussion to the point at which a simple *difference* of right and left articulated in theory (the concept of chirality) is transformed into an observed *asymmetry* (violation of parity), which fixes the meaning of right and left orientation for all eternity. This has important implications for cosmology in ways that Hertz strenuously set out to demonstrate.

CONVENTIONS AND ESSENCE

Shortly after the discovery of the violation of parity in cobalt decay, neutral kaon decays showed that weak interactions violate not only left-right or parity symmetry (P) but also charge-conjugation symmetry (C)—that is, weak interactions distinguish between particles and antiparticles. The composite form, designated as CP violation, implies that laws of physics distinguish between right-handed systems and left-handed systems as well as between particles and antiparticles. The concept of anti-particles had been postulated by the noted physicist, Paul Dirac, as part of relativistic quantum mechanics and, ironically, in spite of Dirac's own skepticism, it has since been experimentally validated that matter has antimatter as its exact counterpart.[46] For every known particle, a corresponding antiparticle exists with the same mass and opposite electric charge. The existence of antimatter, however, raises a fundamental question before the physics community: in the formation of the universe after the Big Bang, why don't we find equal amounts of matter and antimatter? Instead, we find that matter by far exceeds the presence of antimatter in the universe. This CP violation plays a role in explicating the asymmetry of matter and antimatter in the Big Bang narrative: the early universe begins as a uniform, hot, dense gas, which expands and grows exponentially. In this phase of cosmic inflation, all kinds of matter and antimatter pairs are constantly being created and annihilated in equal proportion. As the universe expands, however, it starts getting cooler, and the temperature is no longer high enough to sustain new particle–antiparticle pairs annihilating each other. What emerges in the second just after cosmic inflation are symmetry-breaking, parity-violating processes, which tip the balance toward matter to the virtual exclusion of antimatter. The third phase of cosmogenesis, after matter and energy have taken their present shape, witnesses the gradual emergence of large-scale stable material structures, such as the earliest stars, galaxies, and superclusters.

For the second phase of cosmogenesis, then, a CP violation or the breaking of chiral symmetry forms a key ingredient in explaining the predominance of matter over antimatter.[47] A point of particular contemporary interest, and relevant to our problem of handedness, is the question, how does a particle acquire (certain) mass? Without mass, the universe would be a chaotic sea of particles flying at the speed of light. This is part of the famous Higgs physics—the Higgs mechanism of spontaneous symmetry breaking—that the experiments on the LHC are currently investigating. The "LHCb" experiment at CERN specifically has been measuring the rate of chiral decays at the microscopic level to see how much of it can account for the total amount of matter–antimatter asymmetry observed in the early universe. That is to say, the experiment has tried to reproduce the Big Bang conditions in the laboratory with the aim of figuring out early cosmology. This cosmological explanation, in the form of the Higgs mechanism, is what makes the exploration of handedness so topical. In relativistic quantum mechanics, the problem of mass generation comes down to finding a dynamic mechanism of marriage between right-handed and left-handed constituents of elementary particles. If the Higgs vacuum expectation value is zero, we remain in a chiral and massless world. If the Higgs vacuum expectation value is nonzero, we get an effective right-left interaction, and land in our real, chirally asymmetrical world. To posit the problem on the dynamic origin of mass, some intrinsic difference is needed between massless and massive particles, and this intrinsic difference is mediated by the breaking of symmetry of right and left in electro-weak interactions. Thus, we may venture a harmonizing observation, which does sum up the contemporary experimental situation, that asymmetrical handedness exhibited by point-like particles carries within it the source of and sanction for the origin of the universe.

The harmony, however, conceals a paradox. If dedicated experiments at CERN are probing the amount of CP violation in Strange B mesons or "B-sub-s" decays to explain matter–antimatter asymmetry, it shows that the inner development of the universe can be grasped only as an ineluctable dialectic of microcosm and macrocosm reaching out to each other. I would be justified in concluding that when matter is created, or when the universe comes into existence, it must also be the moment that consciousness takes shape, which is what the asymmetry of right and left intervening in the early process of creation suggests. Raymond Stora, a senior mathematical physicist at CERN, following several long conversations, one day casually observed, "it is unthinkable that the early universe should demonstrate an asymmetrical orientation if there is no absolute direction against which it could be measured." I forthwith replied that if we recall Robert Hertz's formulation that "right and left transcend the limits of our body to embrace the universe," it is no accident that the asymmetry of handedness should play a role in the exegesis of the cosmos.[48] What Hertz shows is that in dual classifications, any one asymmetrical opposition can subsume several others and, furthermore, what can be attributed to the macrocosm applies equally to the microcosm. But these are didactic arguments.

In the more polemical arguments with physicists on this topic, some of them felt that my references to CERN experiments while also questioning physical cosmology into a substrate of anthropological cosmology to be an exaggeration. What they call cosmology is driven entirely by the idea of the universe as a physical entity, approached through the exchange of matter and forces, which answers to quantitative probes. Anthropology, however, takes the opposite standpoint. It advocates a qualitative understanding of cosmology that includes the observer, or rather the community of observers. Indeed, for anthropology, the question of universe is so strongly centered on living creatures of every kind and variety that creation is never a matter of the autonomous development of the physical environment.[49] With such differing points of departure, it is not unusual to find physicists vexed at the suggestion of any crossover between the two. "How can an anthropologist or a philosopher teach a physicist more about space and time?" Wolfgang Lerche, a string theorist and a key informant for this research, once told me pointedly after I acclaimed anthropology for its capacious reflections on nature.

The one-sidedness of their criticism is due in great measure to the fact that contemporary physics, and physicists, cannot conceive of any mediation between the subject and the object of inquiry. Seeds of this mediation, however, are present at the outset in the definition of right- and left-handedness. The failure to appreciate the role of perception (of an observer) and conventions (of a community) in case of particles' orientation is what makes it a matter of bafflement. After the evidence afforded by the experimental discovery of parity violation in beta decay, without hesitation, we can proclaim that physics may try to suppress the observer's presence but cannot abolish it. The subject may be subdued but not eliminated. The core of the problem that asymmetrical handedness represents in contemporary physics is far more radical than is commonly recognized. This problem lies in the fact that the conventions of right and left orientation, which the discourse has attempted to repudiate as extraneous or arbitrary, are now part of the very essence of matter. Handedness, like mass or charge, is an intrinsic property of elementary particles. The drift of this process can be gathered from Wittgenstein's remark offered in the chapter epigraph.[50] In the stamp of his oracular declaration, our conventions are not forced upon us from the outside, nor is our acceptance of them arbitrary; the inexorability that attaches to social or linguistic conventions is simply proof that in relation to the activity whose essence they constitute, is a depth that answers to ontological necessity.[51] But if this isomorphism holds, then the water-tight compartmentalization between symbolic and material, human and physical, and social and natural, which physics forces upon us, ostensibly dissolves. This disclosure does not spring into view from routine conceptualizations of physics. It takes the uncommonly seductive and genuinely interdisciplinary concept of handedness to express the relations between the physical universe, the human mind, and the conventions of a community, which Robert Hertz had first identified.

When I went to CERN, the problem of handedness occupied my mind incessantly. In the first few months of fieldwork, I attempted to discuss it with most physicists, both from theory and experiment, and was aided in this by their infectious enthusiasm. After the implications of left and right asymmetry came up for a provocative discussion with Luis Álvarez-Gaumé, I decided not to pursue this topic so doggedly. I got busy shadowing ATLAS control room meetings (figure 3.2). In the meantime, I assumed he would have forgotten about my interest in chiral orientation. But one day, unexpectedly, he sent an email attaching the picture of René Magritte's astonishing painting, "Not to be Reproduced," with the subject line, parity is not preserved. Note that the canvas depicts a man looking into a mirror that, strangely, reflects his back instead of the front with the eyes, nose, or the mouth. I went back to Álvarez-Gaumé's office, and a congenial conversation ensued. He remarked that no satisfactory explanation has been given for why the mirror image of left-handed neutrinos have not been found in experiments. Trying to grapple with the importance of experimental validation

FIGURE 3.2 The ATLAS control room.

in the physical sciences, I spoke at some length on Karl Popper's falsification thesis. He listened with interest and immediately proceeded to call me a "popperazzi." After this playful comment, he gave an exquisite outline of parity violation and mused that it "relativizes our claim to objectivity." Equally, I said, it universalizes a symbolic classification. The appearance of parity violation can only be an aberration if the assumption is of a gulf between human conventions and metaphysical certainty. Luis Álvarez-Gaumé guardedly acknowledged that the doubts around parity violation are infectious. He then introduced me to a parallel problem, that of magnetic monopoles, which—*if* found—might possibly be a thread to question the subject–object dichotomy commonplace to physics. But that story would have to wait for another chapter.

4

THE CYCLE OF WORK

Modern Industry had therefore itself to take in hand the machine, its characteristic instrument of production, and to construct machines by machines. It was not till it did this, that it built up for itself a fitting technical foundation, and stood on its own feet.

—Karl Marx, *Capital Vol. 1*

In previous chapters, I considered the instrumentation of the Large Hadron Collider (LHC) as a point of departure to clarify the kind of physics undertaken at CERN. In this chapter, I make the collider the focus of attention to examine the division of labor, the temporality of scientific work cycle, and the relationships that form between physics and engineering.[1] It would be impossible to do justice to the complexity of the collider and all its diverse subsystems, like cryogenics and beam dynamics, without provisionally considering the gas explosion that rocked the LHC tunnel in September 2008, which resulted in a suspension of the LHC's operations for several months. The catastrophe led to soul-searching within the physics community to determine whether it had been a technical fault or a human error, an unforeseen accident or a routine failure. Although these concerns suggest the familiar recognition that whatever feats a laboratory may accomplish, it is constituted by limitations, a decisive question meets us in a distinctive form: what constitutes the "normal" work life of a laboratory? Or to put it differently, are breakdowns and crises integral to the cycle of work or exceptional events? This question helps us discern not only the institutional features of a laboratory but also yields an explanation into the distinct phases and tempo of scientific work life. To be sure, temporality in the sense of duration to which people and things are subjected is a basic component of all activity. To isolate the concept as a measure of social norms (of science), however, we must investigate the external marks of temporality (1) in the commissioning and operation of instrumentation, (2) to trace it to technical procedures and principles that govern machine design, and (3) relate these to the institutionalized division of labor among theory, experiment, and instrumentation. Through the September incident that disrupted the collider, we arrive at a representative moment in the life of the laboratory in which the differentiated unity of pure and applied, or theory and practice, is exposed.

THE SEPTEMBER INCIDENT

On the morning of Friday, September 19, 2008, an electrical connection between two magnets failed during a routine circuit test in sectors 3 and 4 of the LHC ring. At the time, a current of 8.7 kilo amperes was being pushed through the superconducting cables. One of the "interconnect splices" linking the cables between two magnets suddenly developed resistance and disintegrated—producing an electrical arc—which punctured the containers of liquid helium that keep the magnets in their 1.9 kelvins (or –271°C) operating temperature. Two tons of helium gas was instantly released in the LHC tunnel and with such a force that a number of magnets broke their anchors to the concrete floor and were displaced and damaged beyond recognition. The release of helium gas tripped the emergency stop, thus switching off all electrical power and services from sectors 3 and 4 of the accelerator.[2]

As soon as preliminary news of the incident trickled in, speculations started on the causes and the extent of damage, the amount of time required for the repairs, and what that meant for the schedule of the collider's operation. In the first few days surrounding the incident, CERN's management released little information, which created discontentment among sections of scientists. An internal mail from Director-General Robert Aymar on September 20, 2008, with the subject "incident in LHC sector 3–4" spoke vaguely of "a large helium leak" and "a faulty electrical connection" between two magnets, which "probably melted at high current leading to mechanical failure." A few physicists grumbled that CERN was a research organization—not a diplomatic or military establishment—and members of the personnel (i.e., the scientists) had a right to know what had caused the incident because their schedules were going to be affected. They argued, how could an academic environment have secrecy? Yet others believed that secrecy was necessary to avoid dampening the general morale, or that the management must have good reasons for withholding details of the damage, which it would disclose at an opportune moment, or more simply, that in an organization so large, everyone cannot know everything that transpires right away.

Soon, however, intense rumors began circulating on the causes and the extent of the damage done to the accelerator. Joseph Masco has spoken insightfully, "secrecy, however, is also wildly productive: it creates not only hierarchies of power and repression, but also unpredictable social effects, including new kinds of desire, fantasy, paranoia, and—above all—gossip."[3] And gossip does not occur in a vacuum, "it is almost always 'plugged in' to social drama," as Victor Turner observes.[4] Soon, conjectures and speculations on what had led to the calamity, what kind of quality tests had been performed on the "interconnect splices," or why were enough spare parts not available to hasten the pace of repairs, prefigured most lunchtime conversations, as my fieldnotes indicate. It was not uncommon in that period to observe the more aloof theoretical physicists striking conversations with accelerator physicists in the cafeteria, who had a better sense of what had befallen the accelerator, in attempts to extract information on the status of repairs than was officially available. Eventually, the director-general's office released a short report to all CERN personnel on the September incident, received via email on October 16, 2008:

Dear Colleagues, We have today issued an analysis of the 19 September incident at the LHC. Investigations have confirmed that cause of the incident was a faulty electrical connection in a region between two of the accelerator's magnets, which resulted in mechanical damage and release of helium from the magnet cold mass into the tunnel. Proper safety procedures were in force, the safety systems performed as expected, and no one was put at risk. Sufficient spare components are in hand to ensure that the LHC is able to restart in 2009, and measures to prevent a similar incident in the future are being put in place. This incident was unforeseen, but I am now confident that we can make the necessary repairs, ensure that a similar incident cannot happen in the future and move forward to achieving our research objectives. The full report is available here.

https://edms.cern.ch/file/973073/1/Report_on_080919_incident _at_LHC__2_.pdf

Best Regards, Robert Aymar[5]

The director-general's message and the report disclosed the full extent of the damage to the LHC and led to a pall of gloom. The "faulty electrical connection," which had induced the catastrophe, was nothing more than poor soldering. "During repair work in the damaged sector, inspection of the joints revealed systematic voids caused by the welding procedure," lamented Mike Lamont.[6] It also became clear to everyone that the otherwise-minor incident was going to cause a major delay in the reoperation of the machine. The reasons for the delay were as follows: twenty-four of the "dipole" magnets and five "quadrupole" damaged magnets needed to be taken out of the tunnel and sent for repairs. The "interconnect splices" had to be inspected in the remaining sectors. Soot and debris had to be cleaned from the beam pipe. New safety systems and enhanced warning systems had to be installed to prevent similar incidents from occurring in the future. The chief reason for the delay, however, was that the damaged sectors 3 and 4 of the LHC would have to be warmed up for the inspections and repairs to take place. Because the LHC is a "superconducting" accelerator and operates at an abnormally low temperature (1.9 kelvin or −271 degrees Celsius), it would take months for the entire twenty-seven-kilometer underground tunnel area to regain room temperature, before which maintenance or repair staff could not enter the tunnel. For a similar fault, not uncommon in a normally conducting accelerator, the repair time would be a mere matter of days. All these factors at the time suggested a minimum of six months' downtime for the LHC operation.

There was little doubt that the engineering constraints, which decided the timeline of the repairs and the renewed operation, were going to affect the physics situation adversely. The key concern animating most lunch-hour discussions on spares and repairs was the question mark placed on the LHC's physics program and what that meant for the prospects of its rival collider, the Tevatron, at Fermi National Accelerator Laboratory, in Illinois, United States. Any setback to the schedule of the LHC implied an immediate advantage to its rival collider. Operating since 1987, the Tevatron was working at peak performance, and the September incident suddenly created a palpable possibility that it could overtake the LHC in staking the first claims to the discovery of the Higgs particle.[7] When news trickled in in March 2009, that the Tevatron had excluded the Higgs mass in the window of 160–170 gigaelectron volts (GeV), it created a ruckus at CERN. Could the Americans snatch the prize of the Higgs right under the noses of the Europeans? The Tevatron had a head start, their technology was stable, and with three inverse femtobarns of collision data collected—the scientific unit that scientists use to count the number of collisions—they could use "it to blow the LHC out of the water," in Albert de Roeck's words. "Coming immediately after the very successful start of LHC operation on 10 September, this is undoubtedly a psychological blow," bemoaned Aymar.

As is evident, the trans-Atlantic competition between colliders is fast and furious. This brings us to a point we have scarcely considered thus far in the course of our observations. We have focused exclusively on the European laboratory, and the LHC. The competition with Tevatron, however, shows that the LHC is not a leap into the void but rather a development and continuity of systematic, ongoing efforts. From this point of view, it becomes clear that a laboratory is best considered within a system of laboratories, and not in isolation as a self-contained unit. This issue is important, especially when we see how relationships—organizationally, scientifically, and politically—between laboratories swing regularly back and forth. It would be safe to say, in fact, that drawing on their mutual relations, every laboratory is able to consolidate its powers and carry forward its mission with unbroken confidence.

Undoubtedly at CERN, with the impetus of outward competition, the thrust toward inner cooperation could not be far behind. Soon enough in the lab, the sense of helplessness gave way to that of urgency. The schedule and strategy of repairing the damaged accelerator took the spotlight. CERN's website began publishing daily updates on the repair status. Here a landscape starts emerging that not only opens a window into the relationship of temporality to the cycle of work but also leads us to ask if a breakdown might contain the possibility of discovering the logical determination of material activity in science. Marx has shown in exemplary fashion that the objectivity of material production does not reside in its materiality but rather in its form, which brings together (nonhuman) nature, labor, and the human community.[8] Hence, it also illustrates that materiality cannot be grasped if it is considered in isolation. The same recognition must be extended to the instrumentation of high-energy physics. Any analysis of the interior world of a laboratory has to be based not only in the inventory of equipment, raw materials, and tools but also on the distribution of work in relation to the material instruments and the human relations mediating work processes. When this process is put into motion, it leads to the constant self-reproduction of a system, even though its actual progress may be checkered or uneven.

MACHINE PARAMETERS

The LHC Project was ratified by the CERN Council, the highest authority of the organization, which includes European member states, in 1994, with approval for the construction of a 14 teraelectron volts (TeV) accelerator coming through in December 1996. The substantial resources required to build the accelerator led to the involvement of nations outside of CERN member states, which included Canada, India, Japan, Russia, and the United States. The Conceptual Design Report, or the "yellow book," specifying the chief parameters of the LHC, such as the superconducting magnets, radiofrequency (RF) and beam feedback, and collimation was published in October 1995.[9] Although the basic design of the yellow book remains unaltered, substantial modifications in engineering and hardware were introduced over the years as accelerator science and technology kept advancing in precision, performance, and complexity.[10] Add to this challenge the need to keep civil engineering costs down, which implied that the LHC "must exactly follow the geometry of LEP," or the predecessor accelerator, the Large Electron Positron Collider.[11] I have emphasized that a laboratory is best considered in a system of laboratories; it would not be out of place to highlight that even in the chronology of accelerator development, we observe a similar feature, which opens up a fresh perspective on the relationship between scientific discoveries and technological innovations.

To expand briefly, in the physics community, the LHC is often referred to as a "discovery" machine, designed to push the energy frontier and gain new insight into the subnuclear world. In contrast, the proposed next-generation collider, the International Linear Collider, is termed as a "precision" machine that would explore in depth and detail what the LHC discovers. The collider at CERN before the LHC, the LEP, had been designed to measure the masses of W and Z particles to a high precision. The W and Z particles were discovered in 1983 at an earlier CERN collider, the Super Proton Synchrotron, or

SPS. The upshot is that the development of accelerator systems shows an alternating logic of expansion and consolidation, or discovery and validation.[12] Mysteries unraveled in one generation of accelerators have been exploited to further probe and seek precision by the next-generation accelerators. The complementarity of this alternating logic reveals the grounds on which CERN developed the LHC, an instrument that is expected to open a window onto new discoveries, right after precision measurements were conducted by the previous accelerator, the LEP.[13] Such a synthesizing continuity in accelerator development is possible only if different laboratories are entwined in mutual relations, which enables us to take an approach to material culture that harks beyond local context. At the same time, we are mindful of contextual features and press forward to the larger issue of what constitutes the total milieu of scientific instrumentation. Perhaps the most conspicuous local factor emerges when we pause over the engineering design of the LHC, which seems uniquely suited to its requirements.

The energy scale is one of the most important distinctive features of the LHC. To achieve rapid acceleration and high energy, the machine uses advanced "radiofrequency cavities," or dynamically changing electric fields at a radio frequency of four hundred megahertz, which "push particles much as ocean waves help surfers gain speed," as Thorsten Wengler from the ATLAS experiment analogically put it, until the protons are whirling around the accelerator at eleven thousand times per second. Although electric fields are used for particle acceleration in the longitudinal plane, magnetic fields are needed for transverse bending, steering, and focusing the particle beams into precise orbits to optimize collisions. Magnetic force is not linear in that it does not operate in the direction of the magnetic field. It acts perpendicular on the beam of particles. This is unlike the electric force, which acts in the direction of the electric field applied. Therefore, the two effects work in two distinct directions: a longitudinal acceleration due to electric fields, and a largely transverse bending of the trajectory of particles due to magnetic fields. Overall, the distinct geometry of forces is responsible for the way particle acceleration takes place. Because magnetic force is perpendicular to the direction of velocity, it can change only the direction of motion—not its magnitude.

Magnets with transverse fields operating in superfluid helium at 1.9 kelvin (−271 degrees Celsius) make up the backbone of the LHC. These are known as "dipole" magnets and are used to deflect or bend the particle beams along the twenty-seven-kilometer circumference. Although particle motion under the influence of dipole magnets is largely stable, it needs extra focusing to force the particles to remain on a "central" trajectory. Magnets used for squeezing or focusing particles in beams closer together to increase the chances of collision are termed "quadrupoles." As the name suggests, a quadrupole has four poles, two north and two south, that are arranged symmetrically around the beam. A single quadrupole focuses in one direction, and defocuses in the orthogonal direction. In total, the eight arcs of the LHC ring have 858 quadrupole magnets and 1,232 dipole magnets, with each of the dipoles having a length of fifteen meters. Interestingly enough, I learned that the fifteen-meter physical limit to the dipole length "was determined by the maximum length allowed by regular transport on European roads."[14] If, however, we are inclined to conclude that local road conditions determine accelerator needs, we are immediately told that the maximum operational field of the LHC was fixed at 8.3 Tesla, "which," writes Lyndon (Lyn) Evans, the LHC project leader, "has its roots in the realm of quantum mechanics rather than in European Union regulations."[15] Both practical and theoretical considerations establish the configurational possibilities of accelerator physics.

At any rate, the chief problem facing the collider during design and construction was the following: the maximum energy attainable in a circular accelerator depends on the product of the bending radius in the dipole magnets and the maximum field strength attainable. To bend two particle beams and generate field strengths in opposite directions, a large area is required. However, the diameter of the underground tunnel—where the collider was to be housed—at 3.8 meters posed a severe challenge. The LEP collider was built in this tunnel, and it was to be dismantled to make way for the LHC. Constrained by the size of the existing tunnel, it was deemed impossible to fit two completely independent rings of the LHC. Because the bending radius was constrained by the size of the tunnel, the aim was to have the magnetic field as high as possible. The problem was daunting. How could two counterrotating proton beams, that is, two separate magnet apertures, with opposite field orientations be confined in the 3.8-meter diameter of the LHC tunnel?[16] The grave challenge of space limitations in the tunnel, and the need to keep capital and operating costs down, led the team of accelerator physicists and engineers spearheaded by Evans to the adoption of a novel "two-in-one magnet" design. The basic idea of a two-in-one magnet system is that windings for two beam channels can be accommodated in a common cold mass cryostat because magnetic flux is circulating in opposite directions in the two channels that can fit in the existing tunnel. At the same time, this makes the magnet structure extraordinarily challenging because the separation of the two beams has to be small enough that they can be coupled both magnetically and mechanically. Given these conditions, the two beam pipes in the LHC are separated by a mere nineteen centimeters inside a common iron yoke (which returns the magnetic field) and the cryostat. While speaking to me, Evans recounted the challenge facing them given that "this is the first time that the two-in-one magnet design has been built," and they had no prior experience to build on. He added,

> The concept of a two-in-one magnet goes back to renowned accelerator physicist, Bob [Robert] Palmer, of Brookhaven National Laboratory. But nobody had used it. . . . In the late 70s we decided to use it. It made perfect sense for the LHC because it is a proton-to-proton collider and so the magnetic field is up in one aperture and down in the other aperture. You couldn't get this to work with a proton-anti-proton collider. The requirement that the fields must be in opposite directions in the two apertures [for a proton–proton collider] ensures that there is no saturation of the central part of the yoke. This simple geometry of flux lines makes possible the exquisite design of two-in-one dipole structure for the LHC.

The description of the unique design of the LHC among accelerators has earned it the epithet, the "Lord of the Rings," among engineers. Engineers also have spoken spiritedly about the far-ranging role of Evans as the person in charge of the LHC for fourteen years, responsible from conception and design through to construction and operation. Striking a more pragmatic tone, Evans has referred to the LHC as a "people's triumph" and has expressed satisfaction that "the machine is working." It was interesting to see that the engineers and accelerator physicists' exclamations accord with human concerns rather than with generalizations of physical nature. That is to say, the proximity of engineers to machine parameters achieved through human labor is not happenstance but utterly reasonable given that it is through tools and instruments they exercise power over external nature. Let us return to the discussion of the catastrophic incident that brought the collider to a standstill.[17]

EVENT AND STRUCTURE

I use the concept of *periodization*, which was developed by Marxist scholars Louis Althusser and Etienne Balibar, to interpret temporality not as a linear evolution of continuous developments but as the transition of structures, built on inner tendencies and contradictions, which shows different levels of social existence in their "relative autonomy."[18] Balibar writes, periodization is "the concept of discontinuity in continuity, the concept which fragments the line of time, thereby finding the possibility of understanding historical phenomena in the framework of an autonomous totality."[19] No shadowy intermediaries are needed to connect time and history because they make contact in social structures. Periodization is then conceived as a state and a development. I attempt to show how periodization lends itself to the exposition of the September incident through which the distinct phases of scientific work are known through the functional unity of theory and practice. In chapters 2 and 3, I invoked the significance of a necessary unity. In contrast, I now argue that the analytic apparatus basic to instrumentation is the functional unity of theory and practice, which pertains to how units and actions cohere or move apart in accordance with an instrumental logic of means and ends. Detailing the nature of this functional unity is of vital importance because it establishes the central conditions on which physics and engineering meet.

Note that the incident in September had occurred just nine days after the gala start-up of the collider. Most of the accelerator physicists and engineers I spoke to during this period averred that the explosion had given them a jolt. It introduced to them the fragility and complexity of the collider. "We always know that there is a lingering possibility that some technical problem, maybe even very small, can threaten operation anytime. . . . the September incident was a critical and a painful lesson but not without useful consequences," shared Gijsbert de Rijk, from the MSC group (Magnets, Superconductors and Cryostats) at CERN, and one of my key informants. More than their disappointment, what the period following the September incident portrayed is that although efficient and successful organization of professional work can accentuate the nature of coordination among various specialized units and groups, a failure or a crisis can manifest how the collectivity disperses and each technical unit settles or coheres to its niche. In saying this, I am not implying a lack of discord between different units or personnel. What I wish to iterate is that during a breakdown, relations between various specialized units can dissolve, with each unit or sector functioning autonomously and fulfilling the task it is responsible for. This is in stark contrast to cooperation, which is generally carried out in the spirit of self-consciousness, and the work done alone is really a legitimate fulfillment of the same impulse, even if it is not obvious to all. In other words, cooperation is sought as beneficial and propitious; it promotes harmony and leads to the completion of tasks in conscious exchange. A crisis, however, when it occurs, can trigger an awareness of the artificiality and inadequacy of community and communication. The sense of exigency, the strain of repairs, and the desire to return to routine, all of which were justified by an immediate purposiveness or a goal in sight, made it necessary for each subculture of physics to function on its own with little connection to the whole or to its neighboring units. This was evident in the period between September 19, 2008, the day the collider stopped, and October 23, 2009, when it resumed operation, and I observed firsthand the relative autonomy of theory, experiment, and instrumentation.

Without exciting new collision data, the theoretical physicists had little to do. Their discussions and seminars were conducted without liveliness. The conveners of the "EP/PP" seminars began organizing talks on issues peripheral to experimental physics such as, "The Evolution of Religious Beliefs" (August 13, 2009) or "The Strange Friendship of Pauli and Jung—When Physics Met Psychology" (December 10, 2009). Their justification was, as Luis Álvarez-Gaumé who happened to be one of the conveners of the seminar series, explained, "Once the LHC starts, the entertainment will stop." Until then, he claimed, they were aware of "killing time." Most of the experimental physicists went back disappointedly to Monte Carlo (or simulation) data. The few days of operation before the incident had offered them a glimpse of the potentiality of the machine, so they seemed particularly affected, haunted, and dejected by the intervening delay. Their chief concerns were the steady accumulation of data by the rival Tevatron collider, the delay in their own and their students' careers, the possibility of other mishaps occurring in a machine so big, and the danger of losing support of funding agencies for future experiments. Several of the physicists spoke of sinking into inertia once again.

In contrast to the theoretical and experimental physicists, were the accelerator physicists and engineers, energetically on their toes, removing, transporting, and repairing the damaged magnets, manually checking more than 150-magnet interconnections in the five warm sectors of the LHC, carrying out cleaning operations in the vacuum chamber in sectors 3 and 4, installing new DN200 relief valves in half of the machine, and reinforcing the support of one hundred main quadrupoles to provide for their better anchoring to the ground. Key to the engineering work in this period was the installation of a new magnet "Quench Protection System" throughout the machine to prevent similar disasters in the future.[20] All this repair and testing work was being carried out under the scrutinizing eyes of the entire organization.[21] During this time, there was no rhetoric, with sentimental overtones of progress and achievement, advocating the harnessing of work toward a common goal. Instead, I was confronted with the sharp division and separation between the demands of physics and engineering. One of the reasons for this division pertained to the schedule of repairs and safety of the machine. The engineering point of view expressed doubts and fears regarding the *actual* operation of the instrument, whereas the physics point of view favored the *potential* of the instrument, the results to be derived (i.e., high-energy collision data) for which the instrument had been assembled. Most experimental physicists seemed in favor of quick repairs and an expeditious rerun of the LHC so that they could have a modicum of collision data to work with to effectively stay in competition with the Tevatron results. The accelerator physicists, in contrast, took a more cautious approach. The great majority of them preferred carrying out exhaustive repairs before recommissioning the LHC to rule out similar incidents occurring in the future. They thought that the physicists, both experimental and theoretical, vastly underestimated the extent of repair work needed on the machine.

In February 2009, the annual "LHC Performance Workshop" was held in Chamonix, France, to assess the schedule of repair and consolidation work on the LHC, which included several experimental and accelerator physicists. As an aside, in general, the annual workshop at Chamonix "provides the teams operating CERN's accelerators a chance to retreat from the hustle and bustle of everyday work at the laboratory and focus on the near- and far-term future of the accelerator complex."[22] The chief question debated at the 2009 workshop was this: at what beam intensity should the LHC be run? Two key alternative scenarios were presented. Scenario one involved the immediate installation of necessary spare parts, the DN200 pressure relief valves, in the four (warm) sectors, and commencing collisions as quickly as possible. This, however, meant that the machine could deliver only limited performance, running in the range of 3.5 TeV energy per beam, instead of the stipulated 7 TeV per beam. The low-energy run would be followed by a yearlong shutdown to conclude the remaining repairs and installations, and moving onto the specified energy level of 7 TeV per beam. Scenario two involved installing relief valves in all eight sectors of the LHC, undertaking diagnostic tests on the whole machine, and going for a delayed start but at full energy of 7 or 8 TeV per beam. The motivations to the two scenarios cut across various issues, such as the following: Was it possible to minimize the impact of the delay and stay in competition with the Tevatron? How could machine safety be weighed against the need for experimental data? Could the LHC be protected against future incidents in a foolproof way?

The alternative positions were thoroughly discussed at the Chamonix workshop. On the whole, experimental physicists favored the first scenario, namely, a limited run with reduced energy in the immediate, followed by a longer shutdown. In contrast, accelerator physicists pleaded for time and raised the importance of comprehensive repairs and checks. They tended to emphasize the "lessons learned" from the September incident and came up with laundry lists of priority issues that needed to be addressed before recommissioning the LHC in any kind of haste.[23] Needless to say in a situation so complex, the lines of fission and fusion were not tightly drawn. Speaking right after the workshop, the director of Accelerators and Technology, Steve Myers, remarked that midway through the workshop he grew skeptical and switched from favoring the second scenario to the first and back again. Significantly, the workshop ended, as Meyers reported, with "no consensus in Chamonix."

The grounds for the debate will emerge more clearly if we bear in mind that physics is not a homogenous ensemble but is marked by steep divisions among theory, experiment, and instrumentation, at the heart of which lies the institutionalized opposition of pure and applied or theory and practice. It is this opposition—and the legitimacy of their respective demands—that makes it possible to identify the rhythms and routines of technoscientific work life, which came to the surface during the crisis. That is, the crisis enables us to see the distinction of pure and applied in structural (meaning) and in functional (effects) terms. As I noted, the time around the September 19 incident cannot be interpreted as a mere unfolding of successive events but rather as an internal drama in which distinct phases are markedly delineated. Before the incident, all of the subcultures of physics exemplified concerted action. During the suspension of the accelerator, the boundaries were accentuated. At this time, many new features to the machine were discovered, including hitherto unforeseen flaws, that could critically disrupt the future operation of the accelerator. According to CERN, "The teams rallied to the challenge of repairing the machine and finding solutions to prevent any reoccurrence of an incident of this kind."[24] Stock was taken of the problems, and some were set aside for later when the LHC would require substantial upgrades for running at optimum energy of 7 TeV per beam.

In all, the repairs and consolidation work took around fourteen months to complete. In 2009, engineering work on repairs, hardware commissioning, and preparations for beams finally came to an end, and physics took the thrust again. On October 23, 2009, particles entered the LHC for the first time since the September incident, and thereafter collisions began. A switch had taken place. For the community as a whole, now physics took the spotlight, with the accelerator retreating to the background. What is of first importance in this progression or movement was not that crises often erupt and are solved but rather the dynamic combination of physics and engineering that confronted this culture as an objective social reality.

TECHNOLOGICAL EFFICIENCY

The relationship observed between engineering and physics would be incomplete if we failed to consider their exchange in the order of daily work. The routine of daily work, however, can all too often pass unperceived during fieldwork, which I can attest to from experience. Moreover, the plethora of activities happening on any given day at CERN could make it difficult to isolate and delimit the meanings of these structural relations. In spite of this, concepts and abstractions have, as already noted, the rigor and the constancy needed to derive structural meanings. To be persuaded of the definitive part played by labor next to nature, and through which we get a glimpse of the human–machine system in physics cast at an industrial scale, I consider a set of contrasting but integrated concepts (i.e., luminosity and production cross section) on which the accelerator works. Luminosity stands for the measure of the number of collisions per unit of area (microbarn) in per unit of time (second). Production cross section refers to the probability of occurrence of particle interactions during collisions. Experimental physicists can invariably be found in discussions involving luminosity and production cross sections. They emphasize that the cross section is always at a given center-of-mass energy, that is, a fixed number dependent on specific physics processes, whereas luminosity is a factor controlled by the parameters of the machine, and thus is dependent on the state of technology.

In this context, another oft-discussed parameter of collider performance is the energy at which particle beams are accelerated. For instance, the LHC accelerates bunches of protons to 7 TeV energies, colliding them head-on 40 million times a second, with each collision generating thousands of new particles at approximately the speed of light. This is a function of energy. The higher the energy, the more violent is the collision. For beam collision purposes, however, energy without luminosity is of little use to experimental physicists. In this sense, energy is closely aligned to but slightly different from luminosity. Gian Francesco Giudice suggestively explains, "Cars traveling at higher speed produce more spectacular crashes, but heavy traffic is needed to produce a sufficiently large number of accidents."[25] Because experimental analyses critically depend on gathering as large a sample as possible of collision data, accelerators are designed to deliver as high a luminosity as they can.[26]

High machine luminosity can be achieved in a number of ways, for instance, by increasing the *number* of particles in the accelerator, such as grouping particles into "bunches," by increasing the *frequency* at which collisions occur, and by *focusing* the beam more tightly by applying strong magnetic fields, especially as it approaches the interaction point at which the particles collide. It is this vital sense of technological efficiency, on which increased luminosity depends, that amplifies the cross section for particle interactions of interest, like for the Higgs boson or supersymmetry particles. I was led to the relationship of production cross section and luminosity while talking to Ian Hinchliffe. If I add that Hinchliffe is a theoretical physicist who heads the Berkeley National Laboratory's participation in the ATLAS experiment, it might strike the readers that already we find a fusion of the three realms of theory, experiment, and instrumentation, and that I am no longer justified in isolating these subcultures as significantly distinct. I would argue, on the contrary, that the lives of personnel can have continuity and overlap, which I will expand on in a moment, but in their professional work and conceptual abstractions, all distinctions survive. During the conversation, Hinchliffe unlocked the constellation of how scientific comprehension works alongside technological efficiency. He said, "You see there is a very small chance of the Higgs [particle] showing up in the first one or two years of data-taking [at the LHC]. But if SUSY [Supersymmetry] exists, it will show up abundantly and right away." I did not grasp immediately why the Higgs boson should take time to "show up" or why SUSY would be abundantly visible.[27] I sought the aid of Andreas Hoecker for an explanation. He elaborated,

The production cross section for a low mass Higgs at 14 TeV energy is 100 picobarns (pb) whereas a low mass SUSY has a cross section of 50–100 picobarns (pb) at 14TeV[energy]. Since we cannot change the production cross sections, which nature has given, we can only work on the luminosity. So here are two scenarios to show you how with a certain configuration of luminosity, discovery takes place. For example, if the cross section is 200 pb, the integrated luminosity sample 1000 pb-1, then we expect to have produced 200,000 Higgs particles. If the integrated luminosity is 10,000 pb-1, that is, 10 times larger, we would expect 10 times more Higgses produced. For the example of 1,000 pb-1, with 200,000 Higgses produced, 1 in 10,000 decays to 2 photons, so we expect in this sample only 20 photon pairs that come from a Higgs signal. Not too many.

We want to make a discovery, right? Can we do that with the above given data sample? A discovery is defined if the significance of a signal is larger than typically 5. Significance is calculated from the number of signal events divided by the fluctuation in the background. The number of signal events is what we determined above, that is, 20 in our example. The background on the figure is larger. Typically, the background is 5 times larger in number of events than the signal, so say 100 events. But what counts is the fluctuation of the background. The statistical fluctuation of the background is the square root of the number of background events, that is, $\sqrt{100} = 10$.

So, for our example of a luminosity of 1,000 pb-1 the significance of the potential for a discovery is $20/\sqrt{100} = 20/10 = 2$. So not 5! We cannot discover the Higgs in this channel with only 1,000 pb-1. What to do? Well, we run the machine longer or improve the instantaneous luminosity with, for instance, more protons in the machine, so that we accumulate a data sample with 10 times more luminosity, that is, 10,000 pb-1. What are the numbers now? The signal is now 200 events. The background is now 1,000 events. The fluctuation on the background is $\sqrt{1,000}$ ~30. So, the significance for the same channel with 10,000 pb-1, instead of 1,000 pb-1, the significance now becomes 200/30, which is about 6.5. Now we are above 5 sigma. We can claim a discovery!

In this discussion with Andreas Hoecker, I absorbed that the cross section of any particle interaction is a physics process determined by nature, so scientists have no choice but to accept it. Luminosity, in contrast, refers to the technological performance of the machine, which falls under human labor and, therefore, can be consistently improved upon. Because experimental analyses depend on "gathering as large a sample as possible," their efforts are to "design an accelerator, which delivers as high a luminosity as we can."[28] This should leave no doubt on the nonoverlapping knowledge claims, and how physics and engineering are reciprocally aligned in a differentiated unity. From the community's perspective, higher machine luminosity could unravel most of the physical mysteries of nature. This buttresses the dominant role of technology as being indispensable to physics discoveries. Yet this pragmatic perspective, if you will, is accompanied by the reflection that the rate of occurrence of physics processes is beyond their grasp. As it turns out, Hinchliffe's views on both the Higgs boson and SUSY have since been confirmed by the evidence from the LHC. There is no latitude of error as far as the corroboration of the Higgs particle is concerned, even though SUSY has not yet been detected. Although the community is under no illusion that the discovery of the Higgs explains or settles everything, it is convinced that the limits of their field are advantageously set by the possibilities of technology. This reliance on human skill and engineering expertise makes particle physics so canonical despite the myriad changes over the years. It is our aim now to see the form in which this technology is harnessed.

THE DIVISION OF LABOR

Everyday conversations on the organizational aspects of CERN often turn on questions of decision-making or bureaucratic authority and control. Yet the most striking feature of a high-energy physics laboratory is the division of labor among theory, experiment, and instrumentation. Speaking on how the division of labor plays out in their field, Patrick Janot, from the CMS experiment, characterizes not only the antagonistic process but also some of the interests and motivations underlying the distribution of tasks, in the following quotation:

> For obscure reasons, physicists—at CERN as everywhere else—like to categorize themselves in different classes, such as theorists, experimentalists, accelerator physicists, technicians . . . Of course, each class is convinced to be dominant over all the others, and that physics is best served by them. Worse! Theorists often think they create physics with their theory, experimentalists often think they discover or invent physics with their experiments, accelerator physicists often think they make it possible with their instruments . . . and they all think that the other categories exist only to provide them with the technicalities they either do not want—or are not able—to work out: long but straightforward calculations, subtle but repetitive measurements, powerful but dirty tools?[29]

Janot has no hesitation in proclaiming that every group of physicists is convinced of its innate superiority and that the more unrewarding tasks are left to others to shoulder. He is not alone in this view. Several informants articulated similar positions on the division of tasks and personnel, and the associated notion that a clear if informal hierarchy is in place in which theoretical physics reigns at the top and experimental physics is a close second, which is followed by accelerator physics and engineering. The administrative workforce is widely perceived as auxiliary staff. Although I received vague answers on how the system devolves, I was gradually able to link their ideas on the overt hierarchy to look beneath to the power that summons it, the operative demarcation of pure and applied, or between those who "think" and "those who use their hands," as one of the engineers at CERN, Francesco Bertinelli, expressed it. While the anti-intellectualism of engineering is greatly exaggerated by Bertinelli, who is himself an astute intellectual, the economy and repetition with which I encountered the articulation of pure and applied suggested to me that it has an intelligible pattern and methodological validity needed to understand the division of labor in high-energy physics.

Marx and Engels argue that "the division of labor only becomes truly such from the moment when a division of material and mental labor appears."[30] This exhortation is a characteristic reminder to think in binary terms. But it also serves to explain rather well how the devolution of tasks takes place, which obliges us to examine the organizational realm more closely. CERN is officially divided into eight departments: physics, information technology, beams, technology, engineering, human resources, finance and procurement, and general infrastructure services. Of these, the preeminent department is physics. It carries out basic scientific research in both theory and experiment. The departments of beams, technology, and engineering are responsible for the hardware of the accelerator, like cryogenics and magnet testing. The other departments, like human resources or finance and procurement, provide infrastructure support and services. Apart from these clear-cut units, myriad gradations can be found in any department, which is where the structural opposition of pure and applied, or theory and practice, is encrusted. For instance, within theoretical physics, those who develop models are widely seen as engaged in "pure theory" and are characteristically distinguished from those who specialize in "computations," or the implementation of algorithms necessary for model-building work, and also from the realm of "phenomenology," in which predicted values from theory are compared with experimental data. Phenomenology and computations are distinctly perceived to be in the zone of application. Likewise, in the experimental physics community, which is numerically by far more preponderant than the theory group, a distinction is commonly observed between those involved with installation, commissioning, and operation of detectors, who are designated as "applied physicists," and those involved with data preparation and analysis, who are termed as "research physicists." Informants made it plain that analyzing collision data is more interesting and valuable than constructing beam pipes or installing cables. Again, in the accelerator sector, it was impressed upon me that TE (i.e., the technology) Department is the leading arm because it involves research and conceptualization of the star project, the LHC, whereas the EN (i.e., engineering) Department simply provides technical coordination and infrastructure support to the accelerator and the various experiments. Coordination, training, and safety were considered "lower-end jobs, or what one does close to retirement," as a magnets engineer, whose name I shall not disclose, said to me with some bitterness. "When one is young, one likes to think about the mysteries of the universe, play with ideas," he remarked wistfully. Conceptualization and development of technology is regarded as more challenging and rigorous than implementation of safety procedures or "finding applications" of a given technology.

The attribution of differential values to mental and manual labor is also conveyed in the anxiety of a number of informants involved on the hardware side of experiments if they could make a transition to more abstruse tasks in the course of their careers. After nineteen years of working on the "end-cap B" of the CMS detector, David Cockerill, who is based at CERN and affiliated with Rutherford Appleton Laboratory, was both relieved and agitated that the detector was completed and sealed. He was relieved because he had contributed in no small measure toward the completion of the detector. At the same time, he was anxious because what would he do now? When I ran into him a year later in the main cafeteria, he said to me jauntily, "I have moved to data analysis . . . not easy, I have done it." During this exchange, Cockerill spoke on the status rankings of theoretical and experimental physics, which, however, stayed in the margins for me because he focused specifically on the importance of data analysis over detector construction. To be sure, making broad generalizations about the distribution of tasks can be a tricky business because exceptions can be found to any classification. It is prudent to add that I was acquainted with quite a few theorists who worked as phenomenologists in the ATLAS experiment, and a bunch of experimentalists who were involved with hands-on accelerator work. Overall, exceptions at the level of personnel and tasks can easily be found. But when we trace the general lines on which the laboratory is organized, some distinctive principles are discernible. One of these principles, whose intellectual force and meaning I have been trying to indicate in this chapter, and which holds the key to understanding the relationship between physics and engineering, is the division and hierarchy between pure and applied.

Having said this much, I hasten to add that the feature of hierarchy is the least of my concerns. I have a rudimentary idea of the hierarchy without anything substantial to offer by way of analysis. My particular interest lies in understanding how the opposition of theory and practice comes to form the heart of the division of labor, as it was refracted through the prism of the September incident. It was clear that talents and tasks are a source, which in the course of time, evolve into a distinction of manual activity or mental activity (i.e., those who work with their hands, and those who use their minds) or multifariously into modeling, theorizing, designing, or manufacturing. More important, however, the value of this distinction lies not only in its functional consequences for the organization but also in what it signifies, more expansively, as a symbol of knowledge. As Jurgen Habermas forcefully argues, "The technical and practical interests of knowledge are not regulators of cognition which have to be eliminated for the sake of the objectivity of knowledge; instead they themselves determine the aspect under which reality is objectified, and can thus be made accessible to experience to begin with."[31] The aspect under which the reality of instrumentation is made accessible is the duality of thought and action, of pure and applied, of theory and practice. The whole matter of the division of labor is simply an elaboration of this structural opposition, as Marx and Engels keenly observed.

Recent studies in the science and technology field have done much to generate the view that without technology, science is of little consequence. The most obvious illustrations of scientific practice transformed by advances in technology emerge from the biomedical sciences.[32] Steven Shapin goes so far as to draw the conclusion that owing to the widespread prevalence of the engineering attitude, the "massive distinction" between science and technology has faded, and consequently the separation between "the role of the scientist and that of the engineer makes less and less sense."[33] Reasons of space make it impossible to deal with his opinion at length, but I can reasonably confirm that in the precincts of CERN, no confusion exists between the roles of a theoretical physicist and a cryogenics engineer. We should, no doubt, be alert to what technology can achieve; however, a sweeping dismissal of the apodictic validity of pure and applied cannot be made without examining it case by case. Perhaps it is the rise of the "practice turn" that leads scholars to willy-nilly focus on material events as these are manifested empirically, and not infrequently, we meet this with an implied disdain for theories, models, and concepts.[34] Far be it from me to assert that such attempts are false or misleading. I want to emphasize only that modern science is and remains the triumph of a method that puts the bifurcation of theory and practice as the upper limits of its action. In the workings of the division of labor, the transactional relations of pure and applied are operative from the start. The plain truth appears to be rather simple: the division of labor among the three subcultures of physics expresses a synthetic rather than a dialectical unity.

Now, Ian Hacking has already discovered that the unity found in experimental science laboratories is often obfuscating and not convincing, which is why he defends the standpoint of "disunity of science." What makes his observation so serious is the realization that the success of high-energy physics cannot be explained through empirical findings, such as the urge to ascribe to contextual factors, ad hoc procedures, or contingent ensembles the direction of its movement. Hacking argues that owing to the intricate order of specializations in model building and technology, material artifacts and phenomena in any laboratory site are produced by a diversity of tools and techniques.[35] In spite of this, he barely dwells on the division of labor, which is the nerve point of technoscientific productivity, to make it an object of focus in his analysis. Instead of division, he attends to diversity (and proliferation). Stable laboratory science arises, Hacking maintains, when theories, materials, and laboratory instruments evolve in such a way that they match each other and are mutually self-vindicating. Such "symbiosis," however, is held to be a contingent feature of contemporary science, which is presented as a motley of local and transitory coordination. In his lengthy study of microphysics, Peter Galison likewise asserts that no grand logic informs the interrelationships of theory, experiment, and instrumentation. The subfields of physics are simply connected by the mutual adjustments of ideas. He has proposed the language analogy—of pidgins and creoles and their "trading zone"—in the recognition of how physicists across subdisciplinary persuasions collaborate. He writes, "two groups can agree on rules of exchange even if they ascribe utterly different significance to the objects being exchanged; they may even disagree on the meaning of the exchange process itself. Nonetheless, the trading partners can hammer out a local coordination, despite vast global differences."[36] In Galison's mode of reasoning, contingency prevails, but not in any drastic way, for it is counterbalanced by "local coordination."

To an extent, such characterizations are well founded. They certainly reflect the everyday, empirical context of science. However, outside of historical contingency or local context, if we still wish to know what relationships outstanding events or everyday practices bear to structural principles of knowledge and classification, an answer is possible, but it is one that must attend to the structural relations of symbol and instrument, concept and object, thought and action. An initial impulse may lead us to the conclusion that science is wholly heterogeneous and without a center. But a sustained reflection discloses—as the discussion revolving around the September 19 incident emphasizes—that the institutionalized opposition of theory and practice forms the source and the sanction of the division of labor prevailing in high-energy physics. Without this recognition, the concrete ethnographic fact involving the September incident, namely, that those branches of physics that deal with theory or analysis had little role to play when the collider was suspended while the engineering work moved round-the-clock, would not make sense. In short, the fundamental objection I have to the claim of "disunity of science" is that it passes over in silence the mechanisms and sources of this disunity, and instead, it is forced to fall back on descriptions, such as symbiosis and contingency, in characterizing scientific activity, which explains nothing, as Marx would urge, of the internal determination of science.[37]

A clear and undeniable link exists between physics and engineering. At the same time, the link is subject to the contradictory demands of mental labor and manual labor. The more closely we scrutinize the instrumentation at CERN, the more we realize the never-separated, always-pulsating set of antagonistic relations of theory and practice, pure and applied, and subject and object, which I set forth somewhat abstractly in chapter 1. This recognition of antagonistic relations may be traced to the wealth of reflections provided by Marxist scholarship on the materiality and symbolism of artifacts, the support of tools and technology to productivity of labor, and the transformation of work processes wrought by machinery. Even then, perhaps, the importance of enduring relationships would not have manifested itself but for the incident in September when coordinated work came to a complete standstill, and I gained a remarkable opportunity: to observe nothing less than the source of rhythms and routines of technoscientific work life. The subsequent repairs and commissioning of the collider also led my attention to the issue of temporality, not as a linear succession of events but as the separation and combination of different stages of the division of labor. In this respect, I have found the viewpoint of local coordination as much too narrow. Nor does the argument of structural disunity (of science) advance beyond the category of contingency. One could say that the instrument constitutes the nub that orchestrates the conjunction and disjunction of the three main subcultures of high-energy physics. Both these aspects of conjunction and separation—of tasks and personnel—belong to the same logic, only manifesting at different phases in the cycle of work. The separation of tasks came to a vivid expression when the collider suffered a breakdown. Conversely, the conjunction of tasks finds expression in the routine and successful order of everyday work.

5

ART, SCIENCE, AND POSTMODERNISM

I remember Menard used to assert that censure and praise were
sentimental operations which had nothing to do with criticism.
—Jorge Luis Borges

I n 1994, Alan Sokal, a mathematical physicist, submitted a
paper to *Social Text*, which drew a link between quantum
physics and postmodernism. The paper was duly published,
at which point it was disclosed by the author to be a meaningless article, "liberally salted with nonsense," and composed
with the aim of debunking postmodern scholarship.[1] In short, it
was a hoax. The provocation to publish this hoax was threefold:
(1) calling attention to the lack of rigor in the social sciences;
(2) protesting against the misuse of scientific and mathematical
jargon in postmodern writings; and (3) reclaiming appreciation
for plain facts and logical coherence over turgid prose and literary excesses.[2] The principal currents of the Sokal controversy
are now behind us. But the original question raised by the hoax
(i.e., How does postmodernism contribute to our understanding of science?) has lost none of its validity.[3] Indeed, Sokal has
complained that few have come forward for a genuine exchange

of viewpoints or have cared to dispute his arguments with tangible evidence.[4] Perhaps the sheer diversity of content between quantum physics and postmodernism is deemed intimidating to invite a discussion. In this final chapter, I closely examine a policy initiative undertaken by CERN to promote an intellectual exchange with the arts to establish how it accords with postmodern conceptions. The account demonstrates what postmodernism brings to light, why the laboratory favors wresting particle physics out of its ivory towers, and how the emerging scientific ethos, both in its motivation and in its effects, can be understood only with the tenets of postmodernism, which builds on a facile *rapprochement* between the sciences and the arts.

My introduction to the arts was unwittingly kindled toward the end of fieldwork, around late 2009, when an initiative was launched by CERN to bring professional artists under its roof to explore the combined creativity of art and science. This unorthodox effort of CERN in hosting a partnership of artists and scientists—unorthodox as CERN is a pure science laboratory—was guided by the premise that art and science are not opposed or unrelated but instead are harmonious and united. The collaborative enterprise envisaged to go beyond a mere token relationship. Part of the policy, labeled "Collide@CERN," involved monetary compensation and institutional recognition for the visiting artists. The laboratory's view was that the artists selected every year to spend up to three months in residence observing physicists at the accelerator complex would have gained a privileged immediacy with nature and science. Equally, the argument put forward from the side of the scientists was that their craft could be enriched by finding in art the rudiments of creativity and spontaneity. Quick with interchange and reciprocity, each side was said to advance toward the synthesis of scientific and aesthetic contemplation of nature, which CERN had identified as the manifest goal of this exercise. This collaborative venture has since gained widespread traction in popular and social media, and it has been touted as the precursor to future initiatives of greater engagement that span art, craft, science, and culture.

What follows is my attempt to answer three key questions: (1) What kind of critical tools have been fashioned to initiate the dialogue between art and science? (2) What scholarly or disciplinary standards are implicit in the exercise? (3) What does the partnership instruct us on the merits of the intellectual division of labor and exchange? In my exploration of the chief tenets of this exchange, I find that the fusion of art and science is not as extensive or as conclusive to the promise of fusion as intended; that epistemological obstacles arise the moment the argument becomes prescriptive; and that, in all of this, postmodernism, which will be defined later, plays more than a modest role in CERN's decision to yoke art and science. The kind of exchange envisioned is not vigorous and contents itself with catchphrases like "kinetic art," "signatures of the invisible," or art and science being so-called kissing cousins. Without doubt, the exercise was undertaken with the intent of shoring up support for Big Science. In fact, the idea that artists should be invited *after* the scientific experts are done setting up the experimental agenda suggests a trendy postmodern sensibility, and one that may be of interest to Alan Sokal, especially because the laboratory's promotion of artworks is sold as a commentary on the dialectical unity of spirit and matter. Not the least of the weaknesses of the proposed exchange is that it adheres to the patent assumption that art inclines to emotion and science to reason, which does nothing to upend the divided foundations on which the classification of knowledge still rests.

Nonetheless, the banality of the attempt is not uninteresting and brings the issue of interdisciplinarity to surface. It generates an immediate awareness of what cutting-edge science- and technology-based art looks like. Echoes of this can be heard in the rise of "sci-art," a term describing artistic projects that seek fusion with science.[5] What also comes to the fore is how physicists' conversations, for the most part, stir up the indignation of a towering discipline like particle physics having to accommodate itself to performing and visual arts, which has vital implications for the subject of popularization of science. Certainly, at moments like this, when stakeholders of a scientific organization strive to square off with "social" interests, they let the cat out of the bag and reveal the wider politicoeconomic nexus in which particle physics operates. At least, from the policy initiative this much is obvious that science is not alienated from everyday pursuits and seeks them with marked enthusiasm. In sum: intellectual reciprocity is both necessary and desirable. Therefore, let us examine the means and methods by which CERN has tried to accomplish this synthesis with art.

PURE SCIENCE AND FINE ART

"CERN has a new cultural policy." With this bold announcement, the *CERN Bulletin* informed its readers of a new policy that represents "the first official framework" for CERN's engagement with the arts.[6] Four activities were identified as its principal features: (1) the launch of an Artist-in-Residence program, (2) the creation of an honorary advisory "Cultural Board for the Arts," (3) the support for outreach through various media events at CERN, and (4) the setting up of a website to provide relevant information to those artists wishing to work at CERN. Essential to the new policy was the inauguration of an international Artist-in-Residence program called "Collide@

CERN—Creative Collisions Between the Arts and Science," which has since 2010 seen artists belonging to various visual and performing art genres coming to CERN every year to observe and work with particle physicists through lectures, workshops, and exhibitions. Ariane Koek, the Communication Group's designated "cultural specialist," has been key to this project. She strongly believes, "the arts and science are kissing cousins. Their practitioners love knowledge and discovering how and why we exist in the world. They just express it in different ways."[7] In the pursuit of building a meaningful collaboration, Koek has been fundraising and building partnerships with external art agencies to spur the Artist-in-Residence program. The question of funding becomes necessary for the invitation to artists to stay for up to three months at CERN and receive an award of 5,000 Swiss francs a month toward living expenses.

Admittedly, collaborations between CERN's scientists and professional artists had been taking place even before the official launch of the "cultural policy." But Koek believes these initiatives were executed in an ad hoc manner. For instance, Josef Kristofoletti, an artist who had painted a part of the Large Hadron Collider (LHC) on a mural, visited CERN in October 2009. But to gain access, he had to cross, in his words, "a labyrinth of red tape and bureaucracy." He also encountered, as he put it emphatically, "Funding limitations. Safety clearances. Legal hurdles. Training courses."[8] To combat this makeshift, informal approach, it was felt that an orderly and transparent system had to be devised. In a short time, a board of experts was set up to sift through applications and select projects in accordance with peer-review procedures. Koek has written, "Artists normally visit because they know someone who works at CERN. So the process isn't very transparent—which is in contrast to the way the rest of CERN works."[9] In addition to the question of remuneration, access to the accelerator complex and, in particular, figuring out how to synchronize the visit of artists with the rhythms of scientific work also had to be determined.

A solution was found in the concept of a working partnership, that is, the pairing of the visiting artist with a scientist or an engineer for a period of three months at CERN. Termed as *Inspiration Partners*, the scientist's job was to act as an in-house guide to the visiting artists, and in the course of weekly meetings and discussions, help familiarize them with the kind of physics undertaken at the laboratory. One such Inspiration Partner happened to be Luis Álvarez-Gaumé, the director of the Theory Division, who jocularly referred to these discussions as "therapy sessions." At the same time, great care was taken to emphasize that the engagement with art does not involve any tampering with the instruments or disrupting the schedule of the experiments. Even the visiting artists affirmed that they did not intend to cause any interruption to everyday scientific work. After all, hanging in the equation was the paramount question of safety of the equipment and associated health risks, especially radiation, because the beams and collisions were already ongoing. It was only mandatory that within the arc of safety guidelines and procedural regulations, the artist and the scientist should collaborate in the premises and come to a joint reflection on the adventures of physics. As much as the desire was to presage a new age with new ideals, signs of conservatism and caution could be seen on both sides from the start and "there were reservations in the minds of some at CERN" as to exactly what a collaboration of art and science would achieve.[10]

An early distinctive feature of this cultural policy had to do with the question of aesthetic representation of physics processes and phenomena. The cultural policy had invoked the idea and activity of the *invisible* in particle physics. The laboratory's higher management appeared convinced that invisible physical forces are striking for the role they play in explanations like the question of dark matter or supersymmetry. Although scientists may feel the mesmerizing power of invisible forces of nature, they must be relentless in seeking rigorous definitions of what they represent. This challenge can nonetheless be given artistic shape, which would be propitious for the evolution of art as well. We may do well to attend to the role of the invisible in particle physics. For instance, the hypothesis of "dark matter," or matter that is invisible but is inferred as existing, is vital for describing anomalies observed in galactic rotation or anisotropies observed in cosmic microwave background. Or consider the postulate of neutrinos advanced by Wolfgang Pauli to explain the principle of energy conservation in beta decay. Pauli theorized that an invisible particle—the nearly massless neutrino—had to be necessarily present in beta decays, which carried away the observed differences between the energy or momentum of the initial and final particles. The concept of invisibility promotes the recognition that the nature of physical objects is not known before it is allegedly named. This concept has a strong parallel in art. Michelangelo Mangano, a theoretical staff physicist at CERN, drew my attention to Salvador Dali's painting titled "Surrealist Composition with Invisible Figures," which depicts the lack of a concrete figure of the human body that is only suggestively indicated by the shape left on the bed. Likewise, Mangano proposed that Picasso's painting "Fruit Bowl, Violin and Bottle," discloses fragments of a newspaper in much the same way as the deflection of radiation is inferred from the motion of galaxies.[11] Hidden forces can await discovery that, he said, may not have yet been formulated but that may be tacitly inferred and known to exist. For Mangano, the tremendous movement between the visible and the invisible opens a window on the deeper kinship of art and physics.

Let us turn to the second dimension of the collaborative endeavor. A cluster of attitudes was said to sustain the dynamic engagement of art and science. Wonderment, enthusiasm, imagination, and passion have been stoutly proclaimed as attitudes common to both vocations, which makes artists and scientists such ideal partners. What was stressed, however, was not the mere recognition that such a parallelism of attitudes exists but that the manner in which these attitudes are manifested in works of art and science also exemplified a concrete historical relation. That is to say, the explanatory weight was placed on what this set of attitudes has achieved historically in the long march of Western civilization as illustrated in renowned works of art and architecture. In a lunchtime conversations with the physicist Michael Doser and cultural specialist Ariane Koek, they offered the illustrations of Leonardo da Vinci's *Last Supper* and Diego Riviera's *Mexican Revolutions* as revealing continuity with CERN's LHC. It is only fitting, Koek claimed, that paintings are made depicting the dazzling power of accelerators and detectors, capturing the majesty, power, and vision of science.[12] In addition, Doser explained that the determination of attitudes extends to the more creative and playful elements in science, which, he thought, are sometimes lost from the view with time but which could be recovered from the backward glance of art. Something like Hegel's famed sentiment about the owl of Minerva taking flight only at sundown was at stake. In Doser's view, the fine arts, with a spirit kindred to pure physics, would be especially conducive to extracting the boldest and liveliest traits of the scientific process and giving them a distinct meaning and identity.

This brings us to the final ingredient of the collaborative venture: to discern and unlock the secrets of nature. That the

arts and the sciences are both well equipped to illuminate variations of growth and decay in natural processes is a stock phrase I
heard a few times in the field. We may recall John Ruskin's mas-
terly formulation on the closeness of art and science in depicting organic activity. Ruskin held that while modernity may have
"desacralized" nature, a level of redemption may be achieved
from the fact that science can portray nature's activity through
active observation while art can pursue it with creative imagination.[13] Shorn of the religious overtones, Ruskin's mode of reasoning was endorsed *tout court* by the cultural policy. It claimed
that art and science express not only nature's grandeur but also
its esoteric inner truths. Properly speaking, for physics, the ultimate reality of the universe is placed in elementary particles of
matter, and its aim is to unveil the laws governing their interaction so that the many-faceted, orderly activity of nature can be
glimpsed. It is at this fundamental level that the soaring union
of artistic sensibility and scientific faculty was firmly urged. And
given the strong similarities in their outlook and interests, such
a cohabitation was deemed mutually sustaining and rewarding.

By opening itself to art, it appears that the accelerator complex is intent on giving a practical expression to the liquidation
of its earlier solipsism. By and large, the physicists involved in
the cultural policy ceased talking of science as a pristine, isolated venture and instead began to talk of comparing science
with other intellectual pursuits, like art, a comparison that would
reveal both similarities and differences. This trait, perhaps more
than others, marks the essential bent of the "new cultural policy"
formalized in August 2010. Needless to say, an enormous range
of possibilities is brought into play. A number of problems also
surface. As I see it, three difficulties arise at the outset. The first
is the assumption informing the usage of the term *culture*. In
the policy, culture is employed as a synonym for higher forms

of creativity, as though culture is the same as being cultured or refined. By such a reckoning, the definition would exclude significant segments of any population as possessing a culture. On the whole, sociologists and anthropologists prefer to describe a vast area of human behavior relating to food, dress, and speech, and even unornamental ones connected with bathing or sleeping, as part of human culture. They would have serious qualms if the fine arts were promoted as a model representation of culture. It is quite likely that the breadth of what the term *culture* encapsulates may have been motivated by Koek's mode of thinking. She has reflected in an article, which defies attempts at paraphrasing, that "in a 10-page pitch to the then Head of Communications at CERN, I argued that in the twenty-first century there is one simple equation: arts + science + technology = culture."[14] From the anthropological perspective, the way Koek valorizes three particular domains as consonant with the wide field of culture makes the scope of this policy unduly myopic. Misunderstandings may crowd around the connotations of culture, and other than a precedent, this meaning cannot be settled, but this temptation to use an equation to create meaning is ensnared in the same cult of scientism that the laboratory ostensibly is seeking to overcome in its engagement with art.

My second objection concerns the engagement of fine arts and the exclusion of industrial arts from this exercise. Undeniably, the fine arts are the most cultivated of all the art forms. But in so far as giving form to matter is concerned, the industrial arts are more propitious than the fine arts, which would be especially relevant, I would think, for a scientific laboratory.[15] Fabrication with diverse tools or materials, that makes us conscious of our labor, seems much more apposite to the spirit of modern experimental physics. This, it is surprising to find that not only is the separation of fine arts and mechanical arts tacitly upheld but also the policy has completely overlooked the latter from its vision. This revealing oversight indicates the underestimation of the applied component in a pure science laboratory, and it serves to only affirm the importance of underlying schools of thought. This sphere of unconscious intellectual presuppositions, which I have highlighted throughout, is the force behind this manifest activity. If practices exist for us to study them, as some scholars urge, we should be left wholly in the dark as to the oversight of applied arts in a laboratory of pure science even when the arts are being courted.[16] The infringement of mechanical arts may not have actual or immediate consequences, but it results no less in narrowing the conception of art.

My third reservation has to do with the overall guest appearance that art gets in this collaborative venture. Although CERN ostensibly claims that its desire to bring art into the ambit of its laboratory is necessary to reveal the grandeur of nature and, most of all, its inner truth, it views artistic greatness consistently from an instrumental point of view. The policy clamors that in the mission to unearth nature's mysteries through particle collisions, artists can join forces with scientists to better represent the dynamics of these processes. But this is misleading. Evincing the representational dimension of art is only one aspect and moreover, wittingly or unwittingly, this approach prioritizes the received ideas of scientific methods in exploring matter, space and, energy. Admittedly, works of art can also be a source of critique or a means of registering protest against scientific forms of describing things.[17] In my view, this one-dimensionality—that art can represent but not critique scientific processes—constitutes the chief weakness of the policy, and obliges it to be in a constantly celebratory mode. Art, however, is not obliged to be sentimental. It is no accident that the cultural policy endlessly harps on Leonardo da Vinci or Filippo Brunelleschi as outstanding creative minds in a turn that is both nostalgic and rearguard. Furthermore, it feels buttressed by this past knowledge to assemble a formula that can pin down how the aesthetic awareness of scientific research should work. One could conclude that this way of looking at things is tantamount to making works of art an accumulation of exhibits and museum pieces that refrains from going beyond the official outlook of scientific theories and terminologies.

Needless to say, CERN is by no means alone in soliciting art to represent the exalted status of science. It would be useful to recall that several decades ago the National Aeronautics and Space Administration, or NASA, had made a similar gesture to invite professional artists as part of its agenda to facilitate wider public outreach and science communication. Around 1962, NASA administrator, James Webb, sought the help of the National Gallery curator, Hereward Lester Cooke, and tasked a few well-known artists with a singular proposal: to capture the emotionally charged epochal moments of space exploration over and above the pitifully humdrum, mechanical portrayal that photographs effect. Cooke voiced the standpoint that "when a launch takes place at Cape Canaveral, Fla., more than 200 cameras record every split second of the activity. . . . But, as [Honoré] Daumier pointed out about a century ago, the camera sees everything and understands nothing. It is the emotional impact, interpretation and hidden significance of these events which lie within the scope of the artist's vision."[18] The outcome was that over the next few years, inspired by the Gemini, Apollo, and Mercury missions, the artists created colorful art works of landscapes around the rocket launching pads, gigantic space shuttles taking off, or white-suited astronauts drifting weightlessly.[19] Since then, sculptors, animation designers, musicians, and filmmakers have regularly partnered with astronauts and engineers of NASA, both informally as well as under paid residency programs, and built an impressive visual archive of the break-throughs in space missions. A few years ago, however, the space agency's artist-in-residence program got mired in a controversy over the question of paying artists' honorarium from federal government funds and, as a consequence, the program was scaled back.[20] This vexing question of using federal money to employ artists in a scientific lab would, henceforth, consign the funding of such collaborative ventures to private grants. An unintended consequence to emerge from the controversy, though, was how the impetus to use art in the furtherance of science so heightened and exacerbated the contradiction between the civic role of science and the time-honored notion of disinterested science. Isn't science dissociated from material interests? And if it becomes comfortable with instrumental objectives, wouldn't all modes of knowledges dominated by utilitarian goals start flourishing? At the root of this perplexity lies the rift between an applied science and a fundamental science. This contrast should give us pause. Once the partisan promotion of science is executed under the imprimatur of artistic legacy, disinterested science cannot perform its designated role in good faith.

Now as a counterargument, it may be objected that no modern laboratory has embodied the pure strain of disinterestedness; that research agendas have always been artfully aligned with defense contracts, patent offices, or venture capital; and that the power of science rests not on its purity but on the efficient communication of its findings. In fact, we may obtain a better understanding of scientific affairs if we stood the argument of disinterestedness on its head and instead took care to learn how laboratories devise engaging strategies to build cross-sectoral exchanges and embellish the outward trappings of their core research. In such a conception, we can assure ourselves that institutionally funded art projects play a dominant role. Not for nothing lies the singularity of, for example, the collaborative partnership with visual arts that the Massachusetts Institute of Technology (MIT) struck when founding the Center for Advanced Visual Studies (CAVS) in 1967. In creed, the CAVS was born from the Bauhaus vision of unifying art, craft, science, and technology to achieve harmonious social regeneration in the modern era. But in substance, "MIT had a vested interest in promoting the arts."[21] Melissa Ragain writes with devastating irony on how the progress of science and technology groped its way to multimedia art and photography to solve what "was a particular problem at the university level, where a presumed humanist outlook broached questions of scholarly ethics and their compatibility with war research."[22] For, given the backdrop of the Cold War, "in 1968 alone, MIT received $108 million from the Pentagon in order to fund both on-campus research in computer technology and classified off-campus projects in surveillance and missile-guidance systems. Urged by student and faculty protest in the late 1960s, the university was under pressure to divert more resources into solving domestic and social issues. The Center was one aspect of this rebranding that simultaneously helped to justify the place of the humanities at MIT."[23] Implicitly shaped by the tendrils of the military-industrial complex, this question of image has played a crucial part in the adoption of aesthetic choices by scientific laboratories. At any rate, it has become increasingly typical for experimental science labs in synthetic biology or artificial intelligence to commission art works and installations to safeguard the credibility of their knowledge against possible attacks.

The issue of image and branding has been resolved somewhat differently by labs in mathematics or theoretical physics. It is true that these disciplines participate in the politicoeconomic issues of the day. They also seek constant public attention for their contributions through audiovisual arts. But it is conceivable that they care little for military advantages. It might be appropriate to mention the recent art and science collaboration undertaken at the Simons Center for Geometry and Physics (SCGP) at Stony Brook, in New York, as an example. Visitors to the SCGP are greeted in the lobby by a two-story-high limestone wall of iconic mathematical equations and diagrams. The entire wall, carved in sixty-nine slabs, features thirty-two equations, including a visual proof of the Pythagorean theorem. It pivots on the uniqueness of mathematical glory. What is not easy to miss, however, is the ironical fact that the element of triumph draws its force from incomprehensibility. To behold the symbols of an equation that withhold immediate comprehension is an acknowledged template in the promotion of scientific awareness.[24] Still more important, because more numerous and routine, is the atmosphere of outreach constituted by "science-inspired art exhibitions," which align with the latent desire to give greater public exposure to scientific expertise.[25] In case there was a doubt, we learn that the SCGP changed the name of its Art and Science Program to the Art and Outreach Program, which expresses the immediate reasoning behind the overture to art.[26] Bearing this in mind, we can say that care is necessary in handling what the expectations are from art and what fulfills them.

Today, it is no longer possible to ignore in the practice of science public relations initiatives that absolutely insist on art installations that promote STEM outreach. It makes science political. Sociologically, this opens a crisis—a twofold crisis, in fact. The first crisis affects the concept of disinterested science, which exercises a palpable hold on most members of the physics community at CERN, for instance, and they begin to question the organization's decision-making that links their enterprise to extraneous political, economic, and ethical concerns. In the abstract, the question takes the form of defending the rights of nature. The second crisis, not unrelated to the first, is that if art best manifests the qualities of science, then what kind of art could or should be used to reflect on modern science, its power, or its autonomy, as well as its limitations? In other words, is it possible to break free from the merely commemorative aspects of art? Considering its prodigious flair for topical relevance, it is not surprising that CERN has deemed a partnership with art to be a worthy pursuit. At the same time, it is perplexing that a more constructive collaboration has not been accomplished, and the result is a kind of bifurcated thinking oscillating between "romantically utopian" and "brazenly instrumental" goals.[27] I next introduce reflections from informants to underscore how in this initiative, cross-disciplinary engagement indeed suffers from an atrophied outlook and demonstrate that the laboratory has not been able to advance to a higher intellectual synthesis to tap into art.

POPULARIZATION OF SCIENCE

I start with an ethnographic vignette. In December 2009, a special lunch was organized in the main cafeteria's "glass box" on the occasion of Arthur I. Miller's visit to CERN.[28] Miller is a historian of science and was visiting CERN to give a talk on "The Strange Friendship of Pauli and Jung—When Physics Met Psychology." About a dozen people were invited to this special lunch meeting. The group included Ariane Koek and a few members from the press office, as well as physicists Luis Álvarez-Gaumé, César Gómez, Mike Lamont, Michael Doser, and Doser's artist wife, Sylvia Wyder. A conversation on some of the similarities and differences between art and science soon ensued. Wild comparisons were made between two prodigious "geniuses" of the twentieth century, Albert Einstein and Pablo Picasso, and how both had simultaneously entered into an extraordinarily productive time around the year 1905 with the former's exposition of special relativity and the latter embarking on the rose period of paintings. "I sometimes wonder," Mike Lamont remarked, "whether simultaneous breakthroughs are really a coincidence." César Gómez answered with a quick reflection on the importance of "the social milieu" and added Bertrand Russell's milestone essay "On Denoting," also published in 1905, to the year's catalog of achievements. The remark went off on a tangent because it dealt with a specific innovation in analytical philosophy. The conversation duly returned to the common elements between art and science, the complementarity of the subjective nuance of art and the systematicity of science, and the great disparity in the prices that paintings fetch in the market as opposed to scientific publications, which are "priceless." When the discussion shifted from the abstract to the specifics of the Artist-in-Residence program, the atmosphere quickly changed from pedantic orations to banter and laughter. "We should have a chef-in-residence. That is more needed for our cafeteria here." "Who will the visiting artists find worthwhile to observe? What will be their muse—particles or people?"

This sort of banter is commonplace during lunch conversations. New ventures, if they are not directly concerned with physics, do not gain easy acceptance and become occasions for the display of wit and repartee, as I had painfully learned. Why the invitation to art? Why not mountaineering? Could art make physicists less boring? Later, when I was walking out from the lunch meeting, Michael Doser was moved to explain to me why he thought works of art depicting scientific processes are a necessary embellishment for the popularization of science. He cited the exorbitant costs of accelerator physics, one that is overwhelmingly dependent on taxpayers' money, as a reminder of how precarious the situation of Big Science is. Historically, it is true that science has tended to be an aristocratic pursuit with "gentlemen" of means and time forming the definitive locus of scientific authority.[29] But today's science is likely carried out by people less aristocratic than their forbears, which is why the question of engaging with the wider public through art occupied center stage in Doser's thoughts. I replied to him that artists are experts in their chosen field, and it might be tricky to aim for popularization with another set of experts. After all, the kind of fine arts that CERN was seeking to rope in had nothing to do with the broad masses. In any event, I explained, greater popularization may not necessarily help the cause of science. I had seen online forums, wikis, and blogs deplore the drain of money in the advancement of accelerator physics when the world gropes with basic survival issues. I argued that many in the public might perceive in the combined efforts of science and art the presence of impractical minds that divert attention from more pressing concerns. Doser objected vehemently and the conversation threatened to take a disagreeable turn, so we stopped.

A few days later, I put the question of popularization of science to César Gómez who had also been present at the special lunch and was familiar with Arthur Miller's work portraying mediations of art, psychology, and science. Gómez grumbled that few ventures have suffered more by popularization than the domain of science: The scientist bemoans the transformation of vocation into a profession, especially as the demand to accommodate public interests are at odds with the rigors of scientific inquiry. The humanist finds in science an intellectual arrogance about truth that simply boils down to a will to power. As for the public, it can swerve any moment against core scientific research, as the opposition to vaccination or mobile phone towers often shows. Gómez went on to expand on the tribulations of impoverishing professionalization, unjustified arrogance, and naïve intellectualism, which in his eyes, was magnified when scientists hitch their wagon to popularization. In contrast, he thought that both physics and art possess an intrinsic dignity. They are much too great to be concerned with purposes, and what is more, they need not be concerned. Gómez voiced his distress at the superficial manner in which particle physics was dallying with the arts as a "field that has lost its way. Hankering after attention is easy when one is laboring under grand illusions." He conceded, however, that nothing could be of greater interest to the ethnographer than the mood that popularization encloses, and more so if art was being pressed in physicists' daily conversations.

For some reason, I avoided clarifying to Gómez at the time that apart from the "glass-box" lunch discussion, the event of the Artist-in-Residence had not elicited widespread response at CERN. Most of the physicists did not care about the artists visiting the complex, and the ones who did could be broadly divided into two camps: the purists and the pragmatists. Since then, I have found these two attitudes to show up consistently over the years, so I can speak with some confidence, on what my interlocutors think of the necessity or desirability of engaging with the arts. From the perspective of the purists, CERN's main mission is scientific and any cross-pollination straight away damages its integrity. "It is ridiculous to send us emails about this or that artist visiting the facility. It has nothing to do with us," a British experimental physicist from the CMS (Compact Muon Solenoid) experiment said irritably when I asked him about the dancers' exhibits. This was the time when dance choreographer Gilles Jobin had come under the auspices of the Artist-in-Residence program and created a piece in CERN's library called "time stands still," which featured three dancers hanging still in various poses from the bookshelves. *The Guardian* newspaper later carried a report that claimed many of the scientists were nonchalant about the performers, "others gave one look and carried on working," and some were "touched" by the dancers.[30] In my opinion, the point is not about the indifference of physicists but that their hair starts to bristle at the intrusion into their spaces. For example, Wolfgang Lerche, a string theorist and a key informant, called out the "trivialization" of science engendered by the cultural policy. He stressed that he was not against art per se, but the "artists are going for the superfluous" by standing in the hallways or in the underground cavern. He was referring to the digital artist Julius von Bismarck, one of the first winners of the Collide@CERN residency award, who had led a team of twenty-two physicists to walk through the LHC's underground tunnel, and on the way recorded their audio impressions of the darkness and chaos as an artwork.[31] Lerche doubted if a single convert to science's cause had been added after these exhibits. In short, the purists protest against the enforced communion with art and defend, with stern pride, the concentrated power of greatness in the autonomy of their science.

In contrast, the second group, the pragmatists, see in the outward accommodation to art a new imaginative energy unleashed for the sciences. This group understands that the engagement with art may or may not be mutually enriching, but it can accomplish something materially for science. Pauline Gagnon, affiliated with CERN's communication group and the ATLAS experiment, was struck by the illustrative value of artwork, which could in the long run facilitate greater acceptance of fundamental physics among the public. She also thought that engaging in active public outreach obliges physicists to be less arrogant and dogmatic. Luis Álvarez-Gaumé, the head of CERN's Theory Division, averred that the arts are pleasant and entertaining and that they could bring recognition to the outcomes of particle physics. To that extent, even with a distinct awareness that the collaboration of art and science is faint or fleeting, why should one forfeit the privilege of infinite publicity? As a side note, Álvarez-Gaumé shrewdly considered that if CERN had other major scientific accomplishments besides the Higgs boson discovery to showcase, it may not have been so driven to seek attention through the arts. James Wells, a theoretical physicist and a key informant, who had been an Inspiration Partner to a visiting artist in 2012, also concurred that bringing artists to the laboratory "is not a catastrophe" because it cannot diminish physics' intrinsic appeal. He was set against the idea that high-energy physics could be content with its own inwardness. In a frank admission, he judged the dangers of an experimental science that is forever dependent on "nature and public money," which can ill afford to lapse into obscurity.

At this point it might be interesting to consider what does all this mean from the laboratory management's point of view? I invite the reader's attention to an article titled "Foundations for the Future," penned by Rolf-Dieter Heuer, CERN's director-general between 2009 and 2015, for its predominantly pragmatic overtones. Heuer notes, "the importance of scientific results are remarkable, but it is the political legacy of the LHC" that he wishes to consider, which includes an affirmation of "basic science" in terms of receiving manifest funding and support from the public, and in turn, inspiring young people to take up scientific careers.[32] The piece acknowledges the advantages of forging diversified partnerships and concludes that success depends on "the right balance of pure and applied research." In short, keenly self-conscious of its leading role, the higher management wants to affirm "how much CERN values its significant role in culture."[33] But, as we have seen, internally, opinion is highly divided on the merits of the collaborative venture, which brings us back to our principal question: What is the common point of departure for the resolution of artistic and scientific harmony? In light of the foregoing remarks, it is tempting to argue that contemporary physics cannot be open to art unless it gives up its self-serving or dilettantish approach and that the whole scheme is rife with compromise. At the same time, this compromise, for one, has deep roots in the intellectual division of labor in which specialized knowledges are vouchsafed their privilege in the guise of autonomy and, two, the tendency toward self-regulation has to battle against external accommodation. In other words, we must extend our analysis to principles of intellectual classification to understand why an outward accommodation of disciplinary genres dominates CERN's milieu. These two aspects are intimately related. But let us examine them one at a time.

SYMMETRY AND AESTHETICS

When the art-science collaborative initiative was launched at CERN, I was in the process of wrapping up my fieldwork. At this time, conversations with informants were not conducted in the form of granular interviews. My association with CERN for of two and a half years also meant that I experienced an inseparable mixture of seriousness and play. Informants, who lived and breathed physics, conveyed many a home truth through means of irony. This held a distinct advantage for the ethnographer—of obtaining "hard data" through the lens of self-reflection. In this context, I should mention an off-the-cuff remark made by a physicist, Massimo Giovannini, that stayed with me for a long while. Observing the art-science venture aloof in the manner of spectators, he said, "there is art but no aesthetic here." I did not have the occasion to ask Giovannini for further clarification on what art is or what he thought its connection to aesthetic involves because I was close to the end of my fieldwork stay at CERN.[34] But the manner in which he uttered it led me to wonder later if scientists have an appreciation of aesthetics from within their work.

For the most part, physicists adhere to a vigorous methodological orientation with an emphasis on observation, deduction, measurement, and analysis. The strength of the method resides not in synthesizing features of every material object in the universe but rather in the identification of boundary conditions, laws, and causes that highlight the regularity of the physical world.[35] In this pursuit, certain aesthetic notions, such as elegance, proportion, and simplicity, play a key role. Regularities observed in natural processes, especially the ones that are acknowledged by dynamic laws, are easily associated with the idea of beauty. This notion is better known as symmetry.[36] As explained in chapter 3, symmetry is the property that defines invariance under certain types of transformation, such as rotation, reflection, or repetition. For example, turning around a tennis ball does not alter the ball's appearance. It is said to have rotational symmetry. Or consider the English letter "H." It possesses reflection symmetry because it looks the same when viewed from the mirror. These are instances of symmetries describing geometrical operations on objects. The strongest argument for symmetry considerations in the arsenal of theoretical physics, however, is when they apply to laws of space and time. Gian Giudice writes, "if mathematics is the language of nature, symmetry is its syntax."[37] It is fair to say that symmetry considerations have profoundly affected the developments of Standard Model physics. Although the Standard Model teaches that all physical reality subsists and exhausts itself within matter and forces, it is guided neither by matter nor forces but rather by the mathematical principle of gauge symmetry.[38] This point is important for our purpose in that the workings of proportion and elegance can be found in the very tissue of subnuclear physics, which suggests the idea of an aesthetic function.

Robert Fleischer, a theoretical physicist specializing in flavor physics and a key informant for this research, carefully explained how mathematical elegance aids in recognizing uniformities in the diversified realm of nature. He considered the principle of symmetry to be a tool no doubt, but a powerful tool, which provides a deeper answer to the *why* questions that are central to modern physics. Fleischer referred to the striking case of Paul Dirac's relativistic equation on the quantum mechanics of particles with spin-half. Based on the symmetry between the properties of positively and negatively charged particles, Dirac predicted the existence of antimatter. The crucial point is that the equation does not change when the sign of the charge is reversed, and from this symmetry Dirac deduced the existence of antimatter, or matter with an opposite sign. As Ian Stewart writes, "Paul Dirac believed that in addition to being mathematical, nature's laws also had to be beautiful. In his mind, beauty and truth were two sides of the same coin, and mathematical beauty gave a strong clue to physical truth."[39] In an important sense, aesthetic conceptions do live on in the sidelines of the community and exert an influence over some of its thinking. "We show a contempt for aesthetics or emotions, but we can mix the logical with the beautiful in [our] theories," remarked Markus Nordberg, of the ATLAS experiment, while outlining what he described as the "elegant geometry" of Kaluza-Klein theory of extra dimensions.[40]

The question of aesthetics holds interest not only for the practicing scientist but also for the historian of science. Thomas Kuhn has examined the complete shift of perspective that art involves and is certainly right to insist that "in the arts, the aesthetic is itself the goal of the work. In the sciences it is, at best a tool . . . seldom an end in itself."[41] Kuhn identifies other revealing differences as well, and yet both art and science are in a position to significantly influence one another. From an epistemological point of view, I believe, it is Kant who first demonstrated that when science discovers that widely divergent phenomena are bound into a compact, unified picture, a recognition of finality of nature arises and, attendant with it, a feeling of aesthetic pleasure. This feeling, he suggests, is not governed by principles of "pure reason," but rather by a more primitive capacity of the soul to experience pleasure in the act of judging objects as beautiful.[42] With this he makes a move that is famously radical and counterintuitive, namely that questions of taste and pleasure, which are normally considered to be arbitrary and subjective, acquire a stamp of universality. Kant recognizes that in no arena of life do we come across greater disagreements between people than in aesthetic matters. Yet he maintains that the experience of beauty is universal, rooted in human judgment. And it is art that plays a vital role in the universalization of taste. Rather than delve into this more deeply, my intent is to highlight that Kant's breakthrough on the feeling of pleasure as a thread linking artistic creation and scientific intuition relies on the middle ground between reason and understanding. More important, nowhere in Kant or in Romanticism later, on which Kant's treatment of aesthetics was to have a profound influence, does art or aesthetics slide exclusively into the province of subjective feelings. The finest romantics avoided giving a one-sided interpretation of art as an emotional outpouring.

The scientists at CERN, in contrast, adhere to the view of art that is the antithesis of science. A great many of them conceptualize art as the spontaneous overflow of creativity and emotions. They cite works of art as products of a creator's unbridled imagination. For them science, however, is based on rules. Its fact gathering is dominated by systematicity. In the daily experience of listening to informants, my impression was that science is not weary of engaging with art. On the surface, they may appear as different as chalk and cheese, but very little is challenging or confrontational in the relationship (although some of the spouses of physicists tried to convince me otherwise, that at home it was *battle royale* between them, their artistic outlooks, and their husbands, steeped in scientific temperaments). The standpoint of even the few physicists who endorse the collaborative venture was articulated around the point that art reverses the way in which science approaches nature. For most of them, science speaks utterly and purely in favor of the object, that is, external physical reality with all its wealth of causal connections, which can be measured and expressed in precise magnitudes. In contrast, art acquires meaning only in relation to the subjective perception of the artist. Art is interlaced through and through with the stamp of the producer, which is why works of art are signed, whereas science must, in the final analysis, remove all traces of the producer from the view.

These observations lead us toward the only plausible—to my mind at least—inference that it is the total heterogeneity of art and science that creates the basis for their association. The key premise of the evolving collaboration is that the experience of art can answer and overcome the alienation of the spirit brought about by mechanical kinds of science. To see the scientific process in all its richness from within becomes the privilege of art. Hence the invitation to professional artists. In this sense, art is meant to complement and complete the potential of science. In other words, the extraneous exploration of nature in physics works well with the manifestation and celebration of its inner pulse contributed by art. This has practical advantages, and not only the benefits of publicity that CERN derives from art, but the formidable social consensus, which has emerged around their complementarity, as a means to make a common cause between the "two cultures."[43] In this mode of reasoning, art is not only the opposite of science but also the ground for its fulfillment. Certainly, the way in which the Artist-in-Residence program has been conceived and executed, points to complementarity rather than conflict, which is why in a span of few months, CERN's Communication group had launched the collaborative effort with total seriousness and success. Readers might not find anything implausible in the quest for fusion between art and science. Today, such interdisciplinarity has even acquired an air of inevitability. Because it is my intention to account for the ecstatic torrents pouring out of the cultural policy, which has created a mild furor at CERN, the analytical wellsprings of the scheme have to be clearly established. This is the point at which postmodernism looms large and aids in making sense of CERN's foray into art.

POSTMODERN CONNECTIONS

I limit myself to one or two remarks on postmodern connections. Roughly speaking, postmodernism is an attempt to describe new sources of meaning by emphasizing myriad combinations of technology, morality, literature, art, and politics. This attempt is backed by the assumption that mediations are an organic and extremely necessary aspect of life, which demand our attention after the advent of modernity. It is in this sense that the prefix "post" is used, that is, coming after modernity both chronologically as well as logically, which implies that it is from the starting point of modernity that the position of postmodernity may be meaningfully deduced. It would be unfair to conceive of postmodernity, however, as merely an add-on to modernity. On the contrary, postmodernism has produced an intense probing of the intellectual, moral, and sociocultural foundations on which modernity stands, which lucidly illustrate the meaning of postmodernism. As Zygmunt Bauman puts it, postmodernism is "modernity conscious of its true nature."[44] In this sense, postmodernity is not a repudiation of modernity, but it should be viewed as its sovereign unfolding and logical conclusion. We may recall at this juncture how postmodernism boasts of piecing together what was arbitrarily divided. For instance, Todd Gitlin sums it up in a sharp quip when he says, "modernism tore up unity and postmodernism has been enjoying the shreds."[45] Hence, perhaps some of postmodernism's avowed belief to act with freedom and spontaneity, take a stab at ambiguity, or revel in disruption. Where modernity tried to lift the world out of its hinges on the lever of rigid and thoroughgoing dualisms, postmodernism tries to overcome it by making fluid and fragile connections. I am inclined to add my conviction, namely, that partial reversals and ephemeral connections do not bring about synthesis or unity. Despite the vengeance it tries to exact on modernist ways of thinking, postmodernism has been reduced, in some ways, to a web of transitory connections. Fundamentally, what it has not resolved is whether meaningful reciprocity can be obtained from episodic crossings-over and combinatorial arrangements.

Be that as it may, the relationship of art and science crafted by CERN expresses the promiscuity central to postmodernism.

A good deal of irony can be found in the fact that scientists of so positivistic a bent like particle physics should yearn for solace from art. That irony dissolves when we examine the logic with which the two extremes of art and science have been yoked. It is one of the contributions of postmodernism to have sharpened our understanding of how modernity is crowded with the logic of dualisms, which are then forfeited with mediations, to counteract the risk of parochialism.[46] Applied to the present situation at CERN, this presupposition of autonomous and distinct spheres drives the collaboration of art and science toward approval and success. Let there be no mistake about it: subjective expressions or emotions do not play a part in the technical determination of experimental work. Confronting this stuffy atmosphere at the mature stage of science, comes the awareness that science can retract its philistinism by retrospectively bringing in art and doubling down on a partnership. But somewhere in this juxtaposition, the whole question of higher forms of synthesis and reflection is tossed aside. Instead, we are left with a formula—art inclines to subjective feeling whereas science favors objectivity and reason—that repeats over and over again in the policy initiative. Consequently, it becomes difficult to pin down any real point of contact between art and science in the cultural policy; nor do pedantically drawn, inflated similarities in the attitudes of art and science produce any synthesis. As the physicists at the "glass box" lunch meeting joked, mountaineering or cooking can also reflect attitudes of zest, passion, and imagination. Should these too find a spot in the laboratory?

Furthermore, we could make the case for postmodernism even in the pragmatic claim that the collaborative attempt is simply a means to receive public approbation. In a joking, self-deprecating vein Michael Doser once said to me, "we are all media whores." True, on the point of publicity and many others, particle physicists can expressly apply themselves to expediency. Besides, he claimed, a little playfulness can do no harm if the "high calling of particle physics," as it exists, is left unscathed. Therefore, one should not forget that for all the claim of playfulness, the artists have been held back from access to experimental sites, and the clear guideline that the preeminence and routine of scientific work must not be affected is maintained unequivocally. Numerous explanations for these guidelines can be suggested but none so poignantly as the logic of mélange that postmodernism represents: art is needed but can never draw close to science, to say nothing of attaining the same professional exactitude in objective thinking, whereas science should acquire a veneer of aesthetic sensibility, but it should do so without diluting its methods or renouncing its privileges. Having made the essential point of this chapter, I add a final word on the vocabulary of the cultural policy, which too seems reminiscent of postmodernism.

A close examination shows not only a disturbing tendency toward excessive hype but also a rampant use of scientific terminology to make stylistic points. We have learned that the program of arts at CERN is called "collide." The visiting artists are given "accelerate" awards. A fusion of lighting installation and dance is labeled "quantum." Ariane Koek, who took it upon herself to formulate the cultural policy, says, "It's about collisions. You can theorize but you don't know exactly what's going to happen in this laboratory of the imagination. There is debris, and things don't fuse."[47] Koek effortlessly enlists the prose of debris and decays from experimental physics for illustrative purposes. Let me cite another example. A rock-music band called Deerhoof visited CERN on the invitation of ATLAS physicist James Beacham. Beacham explains the collaborative venture, "Ex/Noise/CERN," as exploring the unknown. "During Run 2 of the LHC, we're not sure what we'll find—extra Higgs bosons, dark matter, cracks in the Standard Model—and when we brought Deerhoof to CERN, we weren't sure what they'd do in SM-18. But like the best scientists, they were curious, daring and embraced the unknown—with spectacular results."[48] Some readers might opine that I should go easy on these verbal exaggerations, and see them with a bit of poetic license. Doing so, however, would leave the door open for the sort of allusions that Alan Sokal found so revolting. Sokal set out to lampoon the use of technical vocabulary to achieve stylistic effects. He railed against the terrible absence of rigor in the humanities, which he saw as the quintessence of his own field, mathematical physics. In their book, Sokal and Bricmont have exposed with withering disdain the sloppy thinking prevalent in the humanities.[49] They aver that the disgusting mishmash of scientific, mathematical, and literary vocabulary reaches its apogee in postmodernism.

Without doubt, we owe Sokal our gratitude for disclosing the morass of contemporary academics and the dangers of incoherence when the preeminent virtue of epistemology—truth—is set aside under poetic or pragmatic compulsions. But reading CERN's email notifications on the "collide" collaboration over the years, I cannot avoid the feeling that meaningless circumlocution or arid generalizations are not the exclusive province of the humanities. Quirky, farfetched prose is visible in physics, too. Vague analogies are applied indiscriminately. John Ellis, a well-known CERN theoretical physicist, gave a lecture titled, "Answering Gauguin's Questions with the LHC," which appears to be an exercise in free association.[50] If humanistic preaching under the guise of scientific vocabulary feels misplaced and offensive to Sokal and Bricmont, then so does a scientific laboratory proclaiming art to be a hidden source of meaning and solace. After all, nothing would be more pastiche than persuading us to believe that science and art can be complementary after scientists are done setting the agenda, and while steadfastly adhering to the standard assumption of the autonomy of mathematical science, they preach cheery optimism on blurred boundaries and intellectual harmony. Hardly less absurd is the belief that art is a representation of subjective affects and emotions. The proclamations of the cultural policy refuse to admit that art is capable of critique, dissent, or protest. Those in charge of the cultural policy obviously do not seek criticism, for that would contradict their aim of outreach, but any attempt to whittle down the meaning of art and make it more palatable also involves whittling down the possibilities of genuine intellectual exchange. In short, nobody is denying that over and above the mathematical and empirical methods developed by physics, there is room for a more dynamic science that would comprehend the self in relation to nature. If, however, the whole category of *beaux arts* is sought only to be marked as an exception to technical rationality, it is demeaning. If the incidental constellation of art and science is to kickstart the scientific organization's outreach, then I have nothing to say.

The bustle of art and science partnerships has become so sloganesque in official declarations that it is considered a mild blasphemy for physicists to question the unstated premises or the unfulfilled possibilities of what they observe. Speaking three years after the launch of the Arts@CERN initiative, Rolf-Dieter Heuer stridently observed, "the level of heated debate about the so-called 'Two Cultures' is a constant source of bafflement to me. Of course, [the] arts and science are linked. Both are about creativity. Both require technical mastery. And both are about exploring the limits of human potential." Obviously, to what extent the human potential can be admitted in the measurements and proofs of experimental physics has its difficulties. Most physicists know implicitly what privileges are theirs to defend.

They understand that cultural hybridization can be interesting but is also of limited advantage in their science. Before the launch of the cultural policy, the only time an explicit awareness of culture was made was when food was served in Restaurant 1, the main cafeteria on CERN's premises. Every Monday, the dining hall menu featured "kebabs." For the occasion, one of the chefs would don a Turkish hat while dishing out kebab plates to long queues. On Tuesdays, the menu featured "Asian curry." This time a woman sporting a dress with a Chinese collar would stand behind the counter to serve curry. Readers should note that on Wednesdays, Thursdays or Fridays, neither of the two chefs would be serving dishes. Instead the chefs could be found inside the kitchen cooking standard Continental fare, minus the ethnic hat or the collared dress. Without getting embroiled in a debate on the truth value of cultural stereotypes, I only wish to reflect on how regimented the laboratory's intellectual vision is. Scarcely has the imagination worked in more standard ways than in the discovery of culture in customs of food or clothing. Yet this insipid display is not without import. It is fascinating that after forfeiting the significant part of human nature from their working lives, physicists become conscious of it in humble variations of dining and dress. A pure science laboratory that can rarely admit outside intellectual influences in any meaningful way except in the routine of daily commensality shows in the end how onerous it is to divorce matter from spirit.

EPILOGUE

T his book is the result of long research undertaken in the
years when the Large Hadron Collider (LHC) commenced its experimental run of proton-to-proton collisions. At the time, the particle physics community exhibited
a zealous desire to detect not only the Higgs boson but also
other exotic new particles that would usher in physics beyond
the Standard Model. The state-of-the art collider seemed especially propitious to the prospects of fantastic, new discoveries.
In chapter 1, I showed that that this way of judging and seeing things is quite normal and essential. I also noted that we
would do gross injustice to the scientific process if we imagined
that because of several shortcomings in the Standard Model, the
physicists were intent on the production of exotic novel discoveries. The emphasis on constant novelty is so serious that physics demands a new unit of inquiry—problems—that carries the
field forward purely from within itself, through its own initiatives and activities. Such an approach may bring us a long way
toward the social science challenge of identifying the notion
of society or community intrinsic to science. I will not rehash
what I wrote earlier but instead will tell you that nothing happens in physics that does not bear some relation to the creative

powers and limits of its own work. In any event, no society or community can be evaluated by something external to itself. At the same time, the yardstick cannot be an empirical measure, such as the size of experimental collaborations or the amount of funding it commands. Anyone who pays close attention to the content of physics recognizes that it puts the greatest value on possibilities, which are given shape through a rich language composed of mathematical tools and techniques, metaphysical abstractions, and instruments and materials.

At the same time, we are presented with a paradox of sorts that alerts us to the dangers of hunting down physical phenomena to their bare constituents, and reaching exhaustion in positivistic reductionism. One may safely say that the discovery of the Higgs boson turned out to be less exciting and fetched far less applause from the community than what had been anticipated. Experimental measurements conducted in the past decade have unequivocally settled that the Higgs boson discovered at the LHC is exactly as the Standard Model had predicted it. In particular, the identification of the two quantum mechanical properties of the particle, spin and parity, have resoundingly affirmed the predictive power of the Standard Model. Spin, as highlighted in chapter 3, refers to intrinsic angular momentum of particles, and parity demonstrates whether the particle remains the same if the spatial coordinates are flipped, as in a mirror reflection. In these two aspects, the Higgs particle is unlike any other elementary particle in that it has zero spin and even parity. This is a close fit with what the Standard Model had hypothesized. In contrast, the mass of the Higgs boson had not been as precisely forecast by the Standard Model. The challenge of ascertaining the mass of the Higgs boson, for example, as it experimentally emerged from the decay of di-photons, discussed in chapter 2, subsequently established that it was around

125 gigaelectron volts (GeV). Perhaps with this experimental confirmation, a foundation was laid for future research. At the same time, when one considers that experimental physics has delivered on the key goals of Standard Model physics and the community is contemplating the invention of more high-tech instruments, then it corresponds to the functional unity of thought and practice, which was elucidated in chapter 4. In June 2020, CERN approved the construction of a colossal new collider, called the Future Circular Collider (FCC), to be operational and readied by the 2040s to replace the LHC.

In the interim, much electronic ink is being spilled in the physics community discussing how the discipline has run out of steam, when the Standard Model will reach an absolute dead end, or whether the experimental confirmation of the Higgs particle has thrown a spanner in the works for other models and theories. I think that, by itself, this is a mode of reasoning that no one finds prudent or possible to refute. To be sure, any experiment, if it is not mere nonsense, always carries the risk of not revealing manifold variations. But strong, lingering doubts remain as to whether particle physics can progress with its own conceptual resources, keeping its privilege and autonomy intact. The history of science teaches us enough to acknowledge the mirage of mechanistic reductionism and why this method cannot go far enough in exploring the diversity of the universe. Without doubt, the physics community is struggling to keep the discipline topically relevant. When we consider CERN's engagement with art, it also demonstrates why the community is eager to broaden the intellectual ambit of its science. Its reflections on fine art may be taken as part of a joint effort to capture the inexhaustibility of physics. In chapter 5, we reviewed CERN's policy initiative regarding the arts. It expresses rudimentary and confused beginnings. Or rather the engagement with art has not been undertaken with enough attention to questions of method, which could have prevented the rhapsodic turning of intuition into perception, perception into vision, and vision into objects simply put up for display. What makes CERN's cultural policy difficult to digest is the instrumental mindset it has applied— that art can embellish or magnify the powers of science—which passes over in relative silence or does so without expanding our understanding of the presuppositions of science modeled after nature that is calculable and mathematically ordered.

As stated at the beginning of the book, I adopted a double-pronged approach to physics, looking into its presuppositions as well as its practices, while conducting fieldwork. Two and a half years of participant-observation at CERN showed that the presuppositions, in particular the dualisms, of subject and object and theory and practice guided every step of the scientific process. Disclosing themselves as a conceptual scheme, organizing an array of values, and prefiguring every native utterance and action meant that the dualisms selected were reliable signposts for the ethnographic inquiry, which constitute science's very strength and give it its distinctive flavor as modern. However, the process of learning and extrapolating from tacit presuppositions to derive empirical generalizations turned out to be more difficult than I initially anticipated. It was fraught with the risk similar to what offended Wittgenstein: attempting to speak about the unspeakable.[1] Indeed, to maneuver and extricate the core from the proverbial husk of utterances is no simple matter. At any rate, making these presuppositions the explicit focus of research was prompted partly by inclination and partly by necessity.

At the outset, when I had approached Murdock (Gil) Gilchriese, of Lawrence Berkeley National Laboratory, at Berkeley in California, he asked me why I proposed to conduct anthropological fieldwork in a particle physics laboratory. In reply, and wishing to sound audacious, I spoke about my motivation to critique contemporary Western science at the pinnacle of the knowledge hierarchy. Gilchriese seemed unruffled. He forthwith made provision for visa and travel documents, which qualified me to go to CERN. I was aware that the gold standard for an ethnography of science was to "follow the action" and capture the cognitive, normative, and institutional context of a laboratory. I understood that this context could not be confined to the immediate laboratory setting, but rather it involved a wider intellectual configuration of the world. I also knew the importance of studying controversies, which disclose how issues are framed amid divergent opinions in the scientific community. These are valuable signposts from the science and technologies studies (STS) literature. During fieldwork, I found myself doubting whether the gist of STS amounted to an expanded empiricism. It would not be out of place to mention what I truly considered my topic to be: how does a community learn to make predictions of new particles, which are precisely borne out in nature? The whole idea of innovations and discoveries is that these break free from their context and, in doing so, they problematize the very notion of what counts as context. That, at least, is how I realized that we do not have much to rejoice in the notion of situated practices or embedded networks, which are so vigorously put forward by STS scholarship, if one is intent on studying radical discoveries and innovations. Physicists' own concerns on exceeding the immediacy of the present in a manner that preserves the openness of the experimental process should also be mentioned. The scientific creed lives by the maxim that the real is always inflected by possibility.

Now the necessary part on why the focus is fixed on presuppositions: During fieldwork, I heard arguments in abundance with scientists on matters that did not subscribe to positivistic

divisions and separations. Otherwise skeptical of every statement that presents itself as a self-established fact, they vouchsafed the demarcation of subject and object, or pure and
applied, at each turn. If attacked on these, they became livid and
the result was unpleasantness on all sides. Once when I tried
to expound on William James's "Varieties of Religious Experience," combined with my knowledge of growing up in a sacred
cosmos, as a rebuttal to the separation of mind and physical
reality, Luis Álvarez-Gaumé savagely retorted, "Why don't
you go to the witchdoctor next time you are sick?" Overall they
seemed uninhibited, open, and accommodating. Yet they could
not accustom themselves to any idea of an alternative way of
doing science. I was able to understand their position but found
no means to explain mine. I was troubled by their presuppositions and at the same time brooded whether it was an illusion to
think that what has been historically achieved could be undone
by arguing for a mere reversal of assumptions. Most of the disagreements with the physicists at the time on "the great divides"
seemed to lead me to the question: Is it possible to dislodge
them from their high pedestal? What an odd thing really, then,
that from the depths of combat in one-upmanship, enduring
friendships were born.

The fundamental point is simple. Modern physics has successfully deposed the subject from its discourse. We would be
underestimating this point if we were to imagine that their dualistic assumptions could be challenged using controversies or
criticisms, which could be for or against a particular theory or
a piece of experimental evidence, but go no further in restoring
the unity of all knowledges. The reliance of physics on "primary
qualities" in concepts is the most unambiguous and emphatic
avowal that can be found. I suspected, however, at least two rare
exceptions—and this is significant. One is the study of rainbow,

in optics, which my former teacher, Jit Singh Uberoi, had discussed with us. For example, the rainbow can be photographed, which confirms its objective character. At the same time, it is always relative to the distance of the observer, which shows that it is, in part, subjective.[2] The second concept in physics in which the subject makes a surreptitious but grand entry is handedness. To clarify, the concept of handedness is subjective not in the sense of exhibiting physicists' opinions or tastes but in the sense that the assignment of right and left cannot be made without presupposing the perspective of a human observer. I pay homage to the slim volume of Hermann Weyl, *Symmetry*, found and borrowed from the Physics-Astronomy Library at Berkeley, before plunging into fieldwork. Weyl makes the extraordinary revelation that Lorentz invariance, snowflakes, and bathroom tiles can be profitably analyzed by the perspective of proportion, or symmetry. This notion of proportion might sound like a modest one. But what he monumentally illustrates is how to fix the subjective in an everyday, operational concept of physics, like the act of determining right and left in magnetism or polarized light, without which the whole challenge of critiquing science would have evaporated in a haze of lofty and extravagant ambitions. Such an undertaking is also onerous. I did not know then that erecting a monument might involve interring oneself in the foundations first. Perhaps that is how it should be. We are speaking of human realities. But to return to the concept of handedness, what makes it so exceptional is that it refuses to bow to any of the usual separations of symbolic and material, quality and quantity, subject and object, concept and reality, microcosm and macrocosm. This has been grudgingly admitted by most of the physicists in the field.

I recall a series of interviews with Daniel Wyler, a theoretical physicist from the University of Zürich, on parity violation. In the course of these, he once remarked that their discipline routinely raises matters that are complex and multifaceted, but he never thought that a concept that is "not a 100 percent objective" could be detected in particle physics. Wyler also maintained that the classification of right and left chirality belongs to the "inner frame" of physics, by which he meant that it accompanies the study of rotations, boosts, and groups of automorphisms. This is an explication put forward at length in chapter 3. Admittedly, some of the prehistory to the discovery of parity violation, for instance, the fine structure of the hydrogen atom that led to the proposal of spin or the Klein-Gordon equation for a scalar field worked out before Dirac's relativistic equation of the electron, has been left out. I also thought it wise to omit the more contemporary part that obtains in the Cabibbo–Kobayashi–Maskawa (CKM) matrix, and Wolfenstein parametrization, which is relevant to the understanding of CP violation and matter–antimatter asymmetry. Fascinating and absorbing as these topics are, they do not additionally illustrate the points already registered, namely, (1) the fact of a simple difference of right and left transforming into an observed asymmetry, (2) the importance of recognizing preferred orientation in the laws of nature, and (3) the role of parity violation as a necessary ingredient in the formation of mass, and by extension, the universe. We know that on this last point, the LHC experiments are still in the process of accumulating evidence. In that sense, chapter 3 on chirality is a reflection on work in progress. The discovery of asymmetrical handedness is conclusive and here to stay. Moreover, our critique is not drawn from any marginal substrate or "subjugated knowledge," to use Foucault's phrase.[3] Spin, helicity, and CP violation spring from the heart of contemporary physics, and this is enough to suggest that the challenge of interrogating objectivity in science from the core of its concepts has been vindicated.

Chapter 2 on signatures leads us to a similar conclusion, that the observer's judgment far from being extraneous to experimental facts is an essential component, although I must say that the work on handedness was a more difficult circle to square. Unlike handedness, the concept of signature belongs, by and large, to the domain of experimental physics. But in spite of this, the meaning of signature is based on the actively receiving subject, or rather the community of subjects, that is binding in the sense that they bring the synthesizing power of their judgment to the discrimination of background from signal. The demarcation of signals and backgrounds is an indispensable factor in experimental confirmation. And experimental confirmation is the preeminent area in which claims of discoveries appear plausible and achieve realization. The discovery of the Higgs boson from the decay of photons is one illustration of how the community's judgment is so entwined with the notion of a physics signature that the two cannot be separated.

With this recognition of the mind of the scientific subject, three points need stressing. First, how remote it is from the official point of view, and even contradicts it. The objectivist strain reverberates loudly in the scientific mind. It took multiple conversations with David Francis, Andreas Hoecker, David Cockerill-Smith, Thorsten Wengler, and several experimental physicists who are at the top of their game to counteract the point that discoveries in nature are instigated in experiments and waiting to be unveiled is simply untrue. This is another way of saying that objectivity, far from being something intuitively known or materially self-evident, can arise only when it works in conjunction with human judgment within a community. The particle signatures identified in the detectors of the LHC are organically bound up with human thought and perspective. Incidentally, the implications of this recognition are exerting their influence on some of the claims lately coming out of CERN that the Nobel Prize decree of awarding "no more than three" is out of step with the spirit of teamwork found in experimental physics. Experiments are not carried out by "lone geniuses" but rather by gigantic collaborations. Certainly, the desire for the recognition of a Nobel Prize has the most intimate of links with the collaborative nature of scientific work and the vicissitudes of impersonal judgment that animate daily work. The real find, from a sociological point of view, is how far the impersonal relation to nature conjoins with the interpretative act central to signatures.

Second, the overt polemics of deciphering the inner constitution of nature in particle physics signatures displays a remarkable affinity with the medieval doctrine of signatures. I have elucidated that with Paracelsus's and Böhme's predisposition to noncausal or allegorical reasoning, which simultaneously enlarges the scope of human cognitive capabilities, a pronounced mystical tone is struck to naturalism in sixteenth-century Rhineland. In contemporary science, the signature of experimental physics has nothing mystical about it. But in semiotic method, the notion that meaning animates nature that is observable to an alert mind has a revelatory quality. What turns the revelation into reality is that the signature is not merely a representation but also participates in and realizes an event. This argument has been worked out through the case of the di-photon decay of the Higgs boson, in which the photons act as causes *and* signs of the Higgs boson at the same time. In all, I could not have hoped for a better concept than signatures to illustrate the radical departure from causal reasoning in the nub of physics, and in tracing it to a tradition that has roots far deeper than the Enlightenment. We are reliably taught that the signature, as a class of signs that is simultaneously efficacious and expressive, was born in the sixteenth century, in the superposition of semiology, concerned with the constitution of signs, and hermeneutics, involved with the meanings of signs,[4] and then "[disappears] from western science at the end of the eighteenth century."[5] Thus far, only history and philosophy have commented on its extraordinary importance, which I have wished to avail for anthropology from contemporary particle physics.

Third, the discussion of signatures stands in the last instance inseparably connected with that of truth. Beyond obvious aspects, by addressing itself to the majesty of truth, experimental physics shows it is surveying with pleasure the expanse of its territory, without which it would remain, to use Kant's words, "a mere acrobatics of reason." That truth is a basic feature of science's orientation explains why such a premium value is placed in subjecting the wealth of theories, events, and measurements to the most rigorous and critical analysis. Certainly, this is even more the case in experimental physics than in theoretical physics, in which almost every finding receives muted praise but voluble criticism. This brings to mind Einstein's quip that a theory in physics is something nobody believes except the person who formulated it, while an experiment is something everybody believes except the person who conducted it. Let us not forget that what implicates belief and truth in a striking, if dramatized, manner is criticism. At a distance, criticism, which destroys any great work or idea, must always appear unpleasant. But it also implies the conviction that disciplinary engagements are of some consequence. The discourse of signatures is devoted to how physicists think about, say, inferring the momentum of particles, measuring the strength of interactions, clarifying to others the contribution of neutral particles that may not leave a physical trace, or suggesting long-term goals from which assertions about an exotic particle can be made. These expressions of suggestion, assertion, and justification circumscribe the striving toward truth. Put differently, at every turn of eliciting signatures, a judging perspective is at work, which discloses the total splendor of the unity of the mind and the world.

Emboldened by finding the unity of fact and value, and subject and object, I alighted on the realm of instrumentation to seek the unity of theory and practice. As stated in the introduction, the point of departure for most of my reflections was concepts, because their concepts are bound to the thinking of the community. Initially, the concept of stochastic cooling appeared to be a touchstone for approaching the dialectical relations of physics and engineering. Stochastic cooling refers to the method of keeping electrically charged particles in the beam of an accelerator from drifting away from the center. This is done to reduce beam divergence and improve on bunch density and interaction rates. Typically, when accelerated particles deviate from their expected trajectory in the beam, that deviation is measured, and an electric kick or field is applied in like proportion to suppress the deviation and bring the particles back into orbit. The method of stochastic cooling was invented at CERN by Simon van der Meer for which he was awarded the Nobel Prize in Physics. This is one of the rare instances in which the Nobel Prize was awarded for a technological application rather than for work in theoretical or experimental physics. The concept of stochastic cooling is borrowed from the kinetic theory of gases, in which heat is equivalent to disorder, and the term "cooling" is used to denote the reduction of disorder in the beam. The concept posits a spontaneity that is not mapped out by theory. Or so it seemed.

Sustained reflection made it clear, however, that I had been hasty in thinking that the technology of cooling made it possible to conceive of it as a unity with the principles of physics. I missed the fact that genuine practice—in the sense of "praxis"—and not conceived "one-sidedly to the purely practical aspect of technique," which in fact prevents the realization of unity, could be assured only if it touched on the species-life of the community.[6] Without the living relation to the community, the effort seemed like an impoverished feigning of Marxist scholarship that, after all, is the most dedicated and farsighted of the dialectical relations of theory and practice. So it came about that in instrumentation I had failed to achieve in concrete form the necessary unity of theory and practice that I had aimed for. Hopefully, some readers would recognize in this admission that the crystallization of logical oppositions, and their mediations, has not been arrived at from a position of ossified structuralism or from spinning an imaginary web of relations. It is the gift of ethnographic fieldwork that the meals look different from the menu.

The unresolved question of a unified praxis aside, the realm of instrumentation did bring out a significant point, that is, that the division of labor in physics cannot be described solely synchronically by ignoring the cycle of work. A key aspect of the division of labor was exposed at the time of the electrical malfunction, and the resultant helium leak, which brought the collider to a complete halt. Through this crisis, we witnessed how the distinct phases and tempo of work are dominated by the differentiated unity of theory and practice. My justification in illustrating the cycle of work at length is that in the usual discussion of relations between the key subcultures of physics, the division of labor is treated as something of a given, or an empirical, fact. Abstractions that allow for the relations of pure and applied, or manual and mental labor, do not come into the picture. The routine and systematic aspect of the division of labor leaps to the fore from the relations of competition and cooperation prevailing among different laboratories. When CERN's experimental community heatedly disputes the evidence of a new particle in the mass range of 740 GeV emerging from Fermilab, or if American physicists decide to switch to electron-positron collisions for precision measurements of the Higgs boson at the International Linear Collider after CERN pushed for hadron collisions to make the discovery of the Higgs particle at the LHC, it is clear that prior conventions and shared assumptions must be at work or substantive exchange between various labs would not be feasible. Those who are sworn to the "disunity thesis" and try to persuade us that the culture of high-energy physics is widely heterogeneous with a patchwork of values and local coordination at best cannot tell us what points to connections beyond a laboratory. For this reason, it is imperative to remember that any laboratory is best analyzed as existing in a system of laboratories so that it is possible to refer to them in a mediated way, without which the work of physics worldwide would certainly wither.

If I may briefly touch upon one more aspect pertaining to the relations between the subcultures of physics. The LHC is an engineering marvel. The human skill and craftsmanship congealed in the machine is in no sense insignificant. It demands years of precision labor before the machine can deliver on high-performance collisions. Not surprisingly, instrumentation leads to expressions of euphoria, duly captured by the media, such as when the LHC commenced operations in 2010. But there is a controversial side, and one which resists the power of human industry and the complexity of machines themselves. I am referring to the September 2008 explosion following the leak of helium gas in the LHC tunnel. The incident had nothing sinister about it. It was fortuitous that no one was hurt in the gas leak. And yet the incident is a painful reminder that the very same power of human craftsmanship in machinery can act with devastating consequences. In fact, just as the LHC was about to commence operations, regular outlines of doomsday scenarios were surfacing in the print and electronic media. Allusions were made to the lethal role imputed to antimatter in Dan Brown's thriller, *Angels and Demons*, produced at CERN and stolen with the intent of destroying the Vatican City, as described in the novel. Agitated by these reports, I had questioned Michael Doser, who is an antimatter specialist, on the odds of the LHC producing a black hole or antimatter with calamitous consequences for the world. In response, Doser observed sardonically that CERN's annual production of antimatter may at most suffice to light a 250 watt bulb for approximately one minute. Helpful as these languid observations are, the public disquiet over the alliance of physics and engineering, judged to be expedient and mutually beneficent, can by no means be considered settled or as of little concern. Indeed, after the Manhattan project, which saw the careful involvement of 150,000 scientists, engineers, and technicians, we cannot treat questions of autonomy of knowledge and technocracy, the public's trust in science policy, or the bad side of applied science, with condescending arrogance. The implications of these issues, and their explanations, take us to the frontier of how the relationship between expertise, informed citizens, and science policy is conceived.[7] This would probably be of little help without considering some of the political aspects of high-energy physics in practice today, which is beyond the scope of this epilogue. The primacy here, as throughout the book, is on questions of method. I harp on the significance of a methodologically oriented way of thinking about science because a central part of the polemic, designated as science wars, is about method.

Nothing is plainer to apprehend than the internally divided relations binding science and technology, but the awkwardness of our efforts in the social sciences so far has been due to

one-sided critiques. When science studies question the representations found in science, consider the approach of social
constructivism, for example, which raises so many reservations
before it can be satisfied by the materiality of outcomes. But
when it shifts attention to the domain of practices, the merest
presence of humans is satisfactory. No distinction is observed if
the practices involve human agency or actions (will) or human
thought or cognition (mind). In contrast, this fact is of the first
importance in the division of science and technology. We can
rely on the purposiveness of technology to make nature mal-
leable. To behold nature as a pure object, however, we then call
upon physics. Everyone senses this paradox of subject and object
traversing in opposite gears in the realms of thought and practice
and, to wit, nothing is chaotic or capricious about this contradictory unity. It is a self-perpetuating formation perhaps best registered by the German expression *Bildung*. How particle physics
makes use of this paradoxical language to convey its truths on
nature does not seem so mysterious once we acquaint ourselves
with a method that is concerned with contradiction, rather than
simply with difference. And as long as we are given to mere
consideration of observed reality, we cannot grasp relations of
negation or contradiction between subject and object, means
and ends, fact and value, and so on unfolding at various levels.[8]
To seize these relations conjointly is the only way to discern the
substantiality of the mode of existence of modern science.

It may seem too esoteric a point, but the danger of not studying natural science through the prism of conceptual relations,
and instead fetishizing the empirical reflexes of its existence, has
led us to be diffident in raising questions about the content of
scientific work, including its inner truth and falsity. The attention to historically evolving and situationally grounded features
has become the touchstone for STS in recent years. But I have

already indicated its untenability in interrogating innovations and discoveries by showing how problematic the notion of context is. The limitations of political economy–driven arguments are obvious in that money or facilities are not in question when the solitary Standard Model Higgs boson is discovered. In polite words, one cannot find anything reasonable in the amount of money spent and the discovery of the Standard Model Higgs boson. Conversely, enormous advantage is gained by approaching experimental physics, and its seemingly unlimited possibilities of development, from the standpoint of a symbolic classification, which shows how significant aspects persuasively answer to a nondual mode of articulation. In its most general form, the ingenious combinations of subject and object and sign and thing disclosed in the concepts of handedness or signatures invite our attention to those assumptions that scientists need and yet find unable to justify from within the ambit of their discourse. The book is offered as a contribution to some of the subterranean, subversive tendencies residing in science.

We may do well to recall that the mode in which the concepts of signature or handedness have arisen in scientific discourse have little in common with modern tendencies. These are just two examples of the hidden, underground dimension of European thought and practice. I have not provided any systematic theory of pre-Enlightenment categories of thought nor a definite idea of why certain forgotten concepts should arise in modified form in contemporary times. It also cannot be denied that the illustration of concrete cases alone exposes the narrowness of modern science. Even here, the method that determines the stand is more important than the stand itself. At the risk of repetition, the concepts I highlighted of signatures and chiral orientation present a necessary unity, not merely a fortuitous juxtaposition. The postulate of nondualism, invoked throughout the book, implies that what is deduced from the concept must also be shown to empirically accord with something in the phenomenal world. Significantly, it amounts to a doctrine of the mind and its relationship to the world. This must, however, sound like a shocking attestation of universal truths in contemporary times in which instances of relativism or constructivism are frequently offered as effective rejoinders to scientific reductionism. Preaching on the virtues of universals is an awkward business in cultural anthropology. It has also long been beyond the pale of discussion.

As insinuated previously, there is a defense of universals in the work of the French anthropologist Robert Hertz. He has praised the asymmetry of handedness as a common principle of symbolic classification and finds in it an example of the implacable unity of the relative and the absolute, or society and nature, following a single dialectical process. Perhaps the universal is also the central point of Claude Lévi-Strauss's conception of the incest taboo, consecrated in the study of marriage rules and kinship structures, which presses well beyond the limits assigned to cultural modes of reasoning by biological or moral meanings of normality. Similarly, Philippe Descola's analysis of human and nonhuman ontologies is a presentation of an important body of work in which distinctions peculiar to the Western mind have the ground cut from beneath them, so to speak, by the experiences of interiority and physicality present from all over the world. More closely considered, these are crystallizations of Durkheim's insights, who was singularly committed to probing what contributes "to forming the intellect itself" to generate a sociological counterpoint to Kant's epistemology.[9] Yet for all its far-reaching thrust, Durkheim's thought remains an unfinished business because, as Mary Douglas recognizes, it fails to include modern science in its ambit of analysis, a failure that Durkheim does not intend but that he invites with the stiff demands he makes upon anthropology.[10] I do not mention this to reproach his scholarship but simply to see whether we can go further in critiquing contemporary science and technology using Durkheimian categories. The writings of some of these French scholars are overlooked or deemed too arcane for beginners, but these have been extraordinarily helpful for analyzing the conditions of experimentation at CERN. That is to say, these distinguished works hold out the hope that the gap between a natural science and a social science may not be so unbridgeable. After this happy beginning in attempting to draw them closer, I can only say that what made ethnographic fieldwork so remarkable was to be among an intellectually disposed and fair-minded people. The author could wish for nothing more than to have been given this opportunity to be present in their midst at the time of the LHC collisions.

NOTES

INTRODUCTION

1. Robert K. Merton, *The Sociology of Science: Theoretical and Empirical Investigations* (Chicago: University of Chicago Press, 1973); Ludwig Fleck, *The Genesis and Development of a Scientific Fact*, trans. Thaddeus J. Trenn et al (Chicago: University of Chicago Press, 1979).

2. David Kaiser, *How the Hippies Saved Physics: Science, Counterculture, and the Quantum Revival* (New York: Norton, 2011); Sheila Jasanoff, ed., *States of Knowledge: The Co-Production of Science and Social Order* (London: Routledge, 2004); Joseph Masco, *The Nuclear Borderlands: The Manhattan Project in Post–Cold War Mexico* (Princeton, NJ: Princeton University Press, 2013); and Andrew Pickering, *The Mangle of Practice: Time, Agency, and Science* (Chicago: University of Chicago Press, 2010).

3. Ian Hacking, *The Social Construction of What?* (Cambridge, MA: Harvard University Press, 1999), 81.

4. Karl Marx, *Grundrisse. Foundations of the Critique of Political Economy*, trans. Martin Nicolaus (New York: Vintage, 1973), 111.

5. The Standard Model is a theoretical explanation of the interactions between fundamental forces and elementary matter. It is hailed as the crowning achievement of twentieth-century particle physics.

6. Tim Ingold, *Anthropology and/as Education* (Abington: Routledge, 2018).

7. Arpita Roy, "Anthropology as an Experimental Mode of Inquiry," in *Anthropology and Ethnography Are Not Equivalent* (New York: Berghahn, 2021).

8. Pierre Bourdieu, *Outline of a Theory of Practice* (Cambridge: Cambridge University Press, 1977).

9. A remarkable example of an unreal state of affairs being possible is string theory. String theory is a framework of a unified theory that combines quantum physics with gravity. One of its failings is that it can never be proven experimentally since it demands high-energy scales that are inaccessible for any current or foreseeable instrumentation. As a result, opinion in the physics community is sharply divided on whether it even qualifies as a science or is at best a pseudoscience outside the ballpark of physics. Drawing on the sharp-tongued Wolfgang Pauli's description of a poorly written student's paper as "not even wrong," implying that even a wrong idea can be valuable if it leads in a productive direction, Peter Woit in a well-known book mockingly finds string theory as not even wrong. See Peter Woit, *Not Even Wrong: The Failure of String Theory and the Search for Unity in Physical Law* (New York: Basic Books, 2006).

10. Gordon Kane, *Supersymmetry: Unveiling the Ultimate Laws of Nature* (Cambridge, MA: Perseus, 2000), 110.

11. Thomas S. Kuhn, *The Essential Tension: Selected Studies in Scientific Tradition and Change* (Chicago: University of Chicago Press, 1977); and Georges Canguilhem, *A Vital Rationalist: Selected Writings from Georges Canguilhem* (New York: Zone, 2000).

12. René Descartes, *Principles of Philosophy* (Lewiston, NY: E. Mellen, 1988), 19.

13. John Dupré, *The Disorder of Things: Metaphysical Foundations of the Disunity of Science* (Cambridge, MA: Harvard University Press, 1993); and Peter Galison and David J. Stump, *The Disunity of Science: Boundaries, Contexts, and Power* (Stanford, CA: Stanford University Press, 1996).

14. Karin Knorr-Cetina, "Strong Constructivism—from a Sociologist's Point of View: A Personal Addendum to Reply to Sismondo's Paper," *Social Studies of Science* 23 (1993): 556–57.

15. Harry Collins, *Are We All Scientific Experts Now?* (Cambridge: Polity, 2014); Michael M. J. Fischer, *Emergent Forms of Life and the Anthropological Voice* (Durham, NC: Duke University Press, 2003); Ian Hacking,

The Social Construction of What? (Cambridge, MA: Harvard University Press, 1999); Sheila Jasanoff, "Genealogies of STS," *Social Studies of Science* 42 no. 3 (2012): 435–41; and Bruno Latour, *Politics of Nature: How to Bring the Sciences Into Democracy* (Cambridge, MA: Harvard University Press, 2004).

16. Donna Haraway, *Simians, Cyborgs, and Women: The Reinvention of Nature* (New York: Routledge, 1991); Bruno Latour, *We Have Never Been Modern* (Cambridge, MA: Harvard University Press, 1993); and Paul Rabinow, *Anthropos Today: Reflections on Modern Equipment* (Princeton, NJ: Princeton University Press, 2003).

17. Andrew Pickering, ed., *Science as Practice and Culture* (Chicago: Chicago University Press, 1992), 2. In more recent publications he has clarified with illustrations how the analysis of scientific practice calls for a shift from a representational (or cognitive) to a performative (or agential) idiom of evaluation. This argument can be found, among other places, in Andrew Pickering, *The Cybernetic Brain: Sketches of Another Future* (Chicago: University of Chicago Press, 2010).

18. Supersymmetry is motivated by the idea of symmetry of two different categories of particles, "fermions" and "bosons." Fermions are particles of matter while bosons are particles of force carriers. Fermions follow the "Pauli exclusion principle," implying no two fermions can occupy the same quantum state simultaneously. This principle gives matter its solidity or materiality. Bosons, in contrast, cohere together. In spite of this radical difference in the behavior of fermions and bosons, the idea governing Supersymmetry is that fermions can be associated with mirror-like particles of bosons, and vice versa.

19. David Bloor, "Durkheim and Mauss Revisited: Classification and the Sociology of Knowledge," *Studies In History and Philosophy of Science Part A* 13, no. 4 (1982); and David Bloor, *Knowledge and Social Imagery* (Chicago: University of Chicago Press, 1991).

20. Gerald Holton, *Science and Anti-Science* (Cambridge, MA: Harvard University Press, 1993); Steve Fuller, "Can Science Studies be Spoken in a Civil Tongue?," *Social Studies of Science* 24, no. 1 (1994): 143–68; Sandra Harding, *Whose Science? Whose Knowledge?* (Ithaca, NY: Cornell University Press, 1991); and Ullica Segerstråle, *Beyond the Science Wars: The Missing Discourse About Science and Society* (New York: SUNY Press, 2000).

21. It should be added that the title of the book, *Unfinished Nature*, reflects the seduction exercised by Goethe in his observation of nature as "complete, yet ever unfinished." Johann Wolfgang von Goethe, *Goethe's Botanical Writings* (Honolulu: University of Hawaii Press, 1952), 244.

22. According to Newtonian laws of motion, moving particles follow straight, linear paths unless acted upon by force. In the LHC, giant superconducting magnets encircle the beams to bend and deflect the protons—to keep them going in a circular trajectory. Energy loss from *bremsstrahlung* in circular accelerators is a massive problem. On the other hand, one advantage of circular accelerators over linear accelerators is that the ring topology allows continuous acceleration, as particles can transit indefinitely. Another advantage is that a circular accelerator is able to achieve higher acceleration in less space compared to a linear accelerator.

23. The terms *high-energy physics* and *particle physics* have been deployed interchangeably throughout. The difference is of linguistic usage. I have found that the term *particle physics* enjoys greater currency in Europe, but in the United States, the term *high-energy physics* is preferred. I am also aware that factual and numerical figures of instrumentation and personnel at CERN have changed in the decade since I carried out my research. The changes would be of interest to chroniclers of the history of ideas. However, the hardness of numbers is not the terminus of our discussion.

24. Superconductivity is the ability of certain materials to conduct electric current without resistance at low temperatures, which results in producing high magnetic fields. The function of cryogenics, or cooling to low temperatures, is to keep the magnets steady in a superconducting state (i.e., with zero resistance).

25. Max Weber, "Science as a Vocation," *Daedalus* 87, no. 1 (1958): 111–34. On this point, I refer to a matter-of-fact remark of Christopher Llewellyn-Smith, a physicist and former director-general of CERN, according to whom, "when justifying particle physics, it is tempting to invoke spin-offs, such as the World Wide Web which was invented at CERN, but in my opinion they provide a secondary argument and the contribution to knowledge should be put first." C. H. Llewellyn Smith, http://public .web.cern.ch/public/en/about/basicscience3-en.html.

26. Vehicles bearing CD or "Corps Diplomatique" license plates enjoy diplomatic immunity. Generally belonging to vehicles attached to foreign

missions or intergovernmental organizations such as the United Nations, these privileged plates are also given to distinguished scientists and senior staff of CERN. There should be no doubt that as an organization, CERN is self-conscious of being both cosmopolitan and acclaimed. For a more in-depth account of its establishment and operation, see the serialized publication of Armin Hermann et al., *History of CERN: Launching the European Organization for Nuclear Research* (Amsterdam: North Holland Physics Publication, 1987), as well as *History of CERN: Building and Running the Laboratory, 1954–65* (Amsterdam: North Holland Physics Publication, 1990).

1. FINDING THE HIGGS BOSON

1. David Kestenbaum, "What Is Electroweak Symmetry Breaking Anyway?," *FermiNews* 1998.

2. Terry Wyatt, "High-Energy Colliders and the Rise of the Standard Model," *Nature* 448, no. 7151 (2007): 277.

3. John Ellis, "Beyond the Standard Model with the LHC," *Nature* 448, no. 7151 (2007): 298.

4. The term *entered by hand* implies that the numerical values of the parameters are derived from experiment with no theoretical basis to their calculations. An example of this is the inclusion of neutrino masses in various extensions to the Standard Model, which are technically assumed to be massless.

5. An elementary particle is one without a substructure. Because an elementary particle is not made up of smaller particles, it forms one of the basic building blocks of the universe from which all matter is made.

6. The terms novelty and innovation have been used interchangeably, although a subtle distinction does exist between the two. In the context of discovery, the sense of novelty predominates. But when work involves techniques, models, theories, and so on, the concept of innovation is invoked by informants.

7. J. P. Singh Uberoi, *The European Modernity: Science, Truth, and Method* (New Delhi: Oxford University Press, 2002), 95.

8. Uberoi, *The European Modernity*.

9. ATLAS, "Technical Proposal," CERN, 1994, 2.

10. Max Weber, "Science as a Vocation," *Daedalus* 87, no. 1 (1958).

11. Leonard Mlodinow, *Feynman's Rainbow: A Search for Beauty in Physics and in Life* (New York: Warner, 2003), 123.

12. Ian Sample, "Is the Large Hadron Collider Worth Its Massive Price Tag?," *The Guardian*, September 22, 2009, https://www.theguardian.com /science/blog/2009/sep/22/particlephysics-cern.

13. Sample, "Is the Large Hadron Collider Worth Its Massive Price Tag?"

14. C. H. Llewellyn Smith, http://public.web.Cern.Ch/Public/En/About /Basicscience3-En.Html.

15. Hugh Gusterson, *People of the Bomb: Portraits of America's Nuclear Complex* (Minneapolis: University of Minnesota Press, 2004); and Joseph Masco, *The Nuclear Borderlands: The Manhattan Project in Post-Cold War New Mexico* (Princeton, NJ: Princeton University Press, 2006).

16. Gian Francesco Giudice, *A Zeptospace Odyssey: A Journey Into the Physics of the Lhc* (Oxford: Oxford University Press, 2010), 3.

17. J. D. Bjorken, "The Future and Its Alternatives," in www.slac.stanford .edu/. . ./BJ_The%20Future%20and%20Its%20Alternatives.pdf.

18. Steven Weinberg, "Viewpoints on String Theory: Steven Weinberg," *Nova*, July 2003, http://www.pbs.org/wgbh/nova/elegant/view-weinberg .html.

19. Several books document the close connection between high-energy physics and the Cold War, such as Peter Galison and Bruce William Hevly, *Big Science: The Growth of Large-Scale Research* (Stanford, CA: Stanford University Press, 1992); Hugh Gusterson, *Nuclear Rites: A Weapons Laboratory at the End of the Cold War* (Berkeley: University of California Press, 1996); Russell J. Dalton, *Critical Masses: Citizens, Nuclear Weapons Production, and Environmental Destruction in the United States and Russia* (Cambridge, MA: MIT Press, 1999); and Mark Solovey, "Project Camelot and the 1960s Epistemological Revolution: Rethinking the Politics-Patronage-Social Science Nexus," *Social Studies of Science* 31, no. 2 (2001): 171–206.

20. Geoff Brumfiel, "LHC Students Face Data Drought," *Nature* 460, no. 7255 (2009): 558.

21. Rolf Heuer, email message to author, March 9, 2010.

22. Paul Rabinow, *Anthropos Today: Reflections on Modern Equipment* (Princeton, NJ: Princeton University Press, 2003), 5.

23. Marilyn Strathern, *Audit Cultures: Anthropological Studies in Accountability, Ethics, and the Academy* (London: Routledge, 2000); see also Sheila

Jasanoff et al., eds., *Handbook of Science and Technology Studies* (Thousand Oaks, CA: Sage, 1995); and Mario Biagioli, ed., *The Science Studies Reader* (New York: Routledge, 1999).

24. Dominique Pestre, "Commemorative Practices at CERN: Between Physicists' Memories and Historians' Narratives," *Osiris* 14 (1999): 205.

25. Gary Taubes, *Nobel Dreams: Power, Deceit, and the Ultimate Experiment* (New York: Random House, 1986).

26. H. David Politzer, "The Dilemma of Attribution," in *Nobel Lecture*, December 8, 2004. See also Frank Wilczek and Betsy Devine, *Longing for the Harmonies: Themes and Variations from Modern Physics* (New York: Norton, 1987) for a personal account of the discovery of asymptotic freedom.

27. Gaston Bachelard, *The New Scientific Spirit* (Boston: Beacon, 1984).

28. Ian Hacking, *Representing and Intervening: Introductory Topics in the Philosophy of Natural Science* (Cambridge: Cambridge University Press, 1983).

29. Giudice, *A Zeptospace Odyssey*, 30.

30. Giudice, *A Zeptospace Odyssey*, 142.

31. See Rabinow's essay "Artificiality and Enlightenment: From Sociobiology to Biosociality," in *Essays on the Anthropology of Reason* (Princeton, NJ: Princeton University Press, 1996).

32. Burton Feldman, *The Nobel Prize: A History of Genius, Controversy, and Prestige* (New York: Arcade, 2000); and Taubes, *Nobel Dreams*.

33. Bruno Latour, "For David Bloor . . . and Beyond: A Reply to David Bloor's Anti-Latour," *Studies in History and Philosophy of Science*, 30, no. 1 (1998): 123.

34. Bruno Latour, *We Have Never Been Modern* (Cambridge, MA: Harvard University Press, 1993.); and Bruno Latour, "When Things Strike Back: A Possible Contribution of 'Science Studies' to the Social Sciences," *British Journal of Sociology* 51, no. 1 (2000): 107–23.

35. For an interesting discussion on the moniker of "pure science," see Peter Galison, "Ten Problems in History and Philosophy of Science," *ISIS* 99, no. 1 (2008): 111–24.

36. Eugene P. Wigner, "The Unreasonable Effectiveness of Mathematics in the Natural Sciences," *Communications in Pure and Applied Mathematics* 13, no. 1 (1960): 1–14.

37. Wigner, "The Unreasonable Effectiveness of Mathematics," 1.

38. Albert Einstein, "Geometry and Experience," in *An Expanded Form of an Address to the Prussian Academy of Sciences in Berlin* January 27, 1921, https://mathshistory.st-andrews.ac.uk/Extras/Einstein_geometry.

39. P. A. M. Dirac, "The Relation Between Mathematics and Physics," in P. A. M. Dirac, *The Collected Works of P. A. M. Dirac, 1924–1948*, ed. and R. H. Dalitz (Cambridge: Cambridge University Press, 1995); Richard Phillips Feynman, *The Character of Physical Law* (Cambridge, MA: MIT Press, 1965); and Alfred North Whitehead, *Science and the Modern World. Lowell Lectures, 1925* (New York: Macmillan, 1925).

40. Wigner, "The Unreasonable Effectiveness of Mathematics," 9.

41. Ludwig Wittgenstein, *Culture and Value*, trans. Peter Winch (Chicago: University of Chicago Press, 1984), 43e.

42. Philippe Descola, *Beyond Nature and Culture* (Chicago: University of Chicago Press, 2013); and Pierre Bourdieu, *Outline of a Theory of Practice* (Cambridge: Cambridge University Press, 1977).

43. See for instance, David Bloor, "Anti-Latour," *Studies in History and Philosophy of Science* 30, no. 1 (1999): 81–112; Latour, "For David Bloor . . . And Beyond"; Peter Galison, "Ten Problems in History and Philosophy of Science; Ian Hacking, *The Social Construction of What?* (Cambridge, MA: Harvard University Press, 1999); and Andrew Pickering, "The Hunting of the Quark," *Isis* 72, no. 2 (1981): 216–36.

44. Sharon Traweek, *Beamtimes and Lifetimes: The World of High Energy Physicists* (Cambridge, MA: Harvard University Press, 1988).

45. Peter Galison, *Image and Logic: A Material Culture of Microphysics* (Chicago: University of Chicago Press, 1997).

46. Trevor J. Pinch, "Opening Black Boxes: Science, Technology and Society," *Social Studies of Science* 22, no. 3 (1992): 487–510.

47. Jester, "*UTfit longo magis quam acri bello,*" *Résonaances* (blog), March 25, 2008, http://resonaances.blogspot.com/2008/03/ut-fit-longo-magis-quam -acri-bello.html

48. The significance of 5 sigma deviations in experimental discoveries will be illustrated in chapter 2.

49. The audience in the Main Auditorium was composed mainly of experimentalists and, therefore in some sense, the skepticism accorded to UTfit's data was higher.

50. In his paper, Michelangelo Mangano, from CERN's Theory Division, cautions the community that a discrepancy or deviation from

Standard Model values by itself should not be treated as a sign of New Physics. See Michelangelo Mangano, "Understanding the Standard Model, as a Bridge to the Discovery of New Phenomena at the Lhc," *arXiv:0802.0026v2 [hep-ph]* (2008): 2.

51. H. M. Collins, *Gravity's Shadow: The Search for Gravitational Waves* (Chicago: University of Chicago Press, 2004).

52. Karl Marx and Friedrich Engels, *Economic and Philosophic Manuscripts of 1844*, Great Books in Philosophy (Amherst, NY: Prometheus, 1988), 105. Italics in original.

53. CERN, http://outreach.web.cern.ch/outreach/

2. NATURE AND SIGNATURE

This chapter was published earlier as "Ethnography and Theory of the Signature in Physics," *Cultural Anthropology* 29, no. 3 (2014): 479–502. I am grateful to the journal for permitting me to reuse the article here.

1. Ernst Cassirer, *The Philosophy of Symbolic Forms*, vol. 3 (New Haven, CT: Yale University Press, 1957).

2. Eben Kirksey and Stefan Helmreich, "The Emergence of Multispecies Ethnography," *Cultural Anthropology* 25, no. 4 (2010): 545–76.

3. Donna Haraway, *Simians, Cyborgs, and Women: The Reinvention of Nature* (New York: Routledge, 1991); Bruno Latour, *We Have Never Been Modern* (Cambridge, MA: Harvard University Press, 1993); Paul Rabinow, *Anthropos Today: Reflections on Modern Equipment* (Princeton, NJ: Princeton University Press, 2003).

4. Sarah Franklin, "Re-thinking Nature–Culture: Anthropology and the New Genetics," *Anthropological Theory* 3, no. 1 (2003): 65–85.

5. Michael M. J. Fischer, *Emergent Forms of Life and the Anthropological Voice* (Durham, NC: Duke University Press, 2003).

6. Paul Rabinow and Gaymon Bennett, *Designing Human Practices: An Experiment with Human Practices* (Chicago: University of Chicago Press, 2012), 7.

7. Hans-Jörg Rheinberger, *Toward a History of Epistemic Things: Synthesizing Proteins in the Test Tube* (Stanford, CA: Stanford University Press, 1997), 140.

8. Donna Haraway, *Modest_Witness@Second_Millennium.FemaleMan©_Meets_OncoMouse™: Feminism and Technoscience* (New York: Routledge, 1997), 113.

9. Marilyn Strathern, *Partial Connections* (Savage, MD: Rowman & Littlefield, 1991).

10. Émile Durkheim, *The Elementary Forms of Religious Life* (New York: Free Press, 1995).

11. Émile Durkheim and Marcel Mauss, *Primitive Classification* (Chicago: University of Chicago Press, 1963), 81–82.

12. Giorgio Agamben, *The Signature of All Things: On Method*, trans. Luca detrans D'Isanto and Kevin Attell (New York: Zone, 2009), 68.

13. Andrew Pickering, *Constructing Quarks: A Sociological History of Particle Physics* (Chicago: University of Chicago Press, 1984).

14. Steven Weinberg, *Facing Up: Science and Its Cultural Adversaries* (Cambridge, MA: Harvard University Press, 2001).

15. Michelangelo Mangano, "Understanding the Standard Model, as a Bridge to the Discovery of New Phenomena at the LHC," *arXiv:0802 .0026v2 [hep-ph]* (2008): 3.

16. This is the notational form specifying the decay of a Higgs into two pairs of leptons mediated by a Z boson and a virtual Z particle (with a mass less than 90 GeV).

17. ATLAS, "Technical Proposal," CERN, 1994, 217.

18. Unlike hadrons, electrons and positrons are elementary particles and lead to relatively clean collisions.

19. Karen Michelle Barad, *Meeting the Universe Halfway: Quantum Physics and the Entanglement of Matter and Meaning* (Durham, NC: Duke University Press, 2006); and Karin Knorr-Cetina, *Epistemic Cultures: How the Sciences Make Knowledge* (Cambridge, MA: Harvard University Press, 1999).

20. Donna Haraway, *Modest_Witness@Second_Millennium*; and Strathern, *Partial Connections|*.

21. Mary Douglas, *Implicit Meanings: Selected Essays in Anthropology* (London: Routledge, 1999), 252.

22. H. M. Collins, *Gravity's Shadow: The Search for Gravitational Waves* (Chicago: University of Chicago Press, 2004).

23. Jacques Derrida, *Margins of Philosophy* (Chicago: University of Chicago Press, 1982).

24. Derrida, *Margins of Philosophy*, 328.

25. Peter Galison, *Image and Logic: A Material Culture of Microphysics* (Chicago: University of Chicago Press, 1997); Mangano, "Understanding the Standard Model."

26. Ferdinand de Saussure, *Course in General Linguistics* (London: Duckworth, 1983).

27. Charles S. Peirce, *Writings of Charles S. Peirce: A Chronological Edition*, 8 vols., compiled by the editors of the Peirce Edition Project (Bloomington: Indiana University Press, 1982–2009); and Tzvetan Todorov, *Theories of the Symbol* (Ithaca, NY: Cornell University Press, 1982).

28. Peirce, *Writings of Charles S. Peirce*; and Todorov, *Theories of the Symbol*.

29. de Saussure, *Course in General Linguistics*.

30. Émile Benveniste, *Problems in General Linguistics* (Coral Gables, FL: University of Miami Press, 1971).

31. Umberto Eco, *Semiotics and the Philosophy of Language* (Bloomington: Indiana University Press, 1984); Giovanni Manetti, *Theories of the Sign in Classical Antiquity* (Bloomington: Indiana University Press, 1993); and Todorov, *Theories of the Symbol*.

32. Michel Foucault, *The Order of Things: An Archaeology of the Human Sciences* (New York: Vintage, 1970).

33. Winfried Nöth, *Handbook of Semiotics* (Bloomington: Indiana University Press, 1990); and Thomas A. Sebeok, *Contributions to the Doctrine of Signs* (Bloomington: Indiana University Press, 1976).

34. Umberto Eco, *A Theory of Semiotics* (Bloomington: Indiana University Press, 1976), 217.

35. Paracelsus, *The Hermetic and Alchemical Writings of Aureolus Philippus Theophrastus Bombast, of Hohenheim, called Paracelsus the Great*, trans. A. E. Waite. New Hyde Park, NY: University Books, 1967).

36. J. P. Singh Uberoi, *The European Modernity: Science, Truth, and Method* (New Delhi: Oxford University Press, 2002), 14.

37. Andrew Weeks, *Paracelsus: Speculative Theory and the Crisis of the Early Reformation* (Albany: State University of New York Press, 1997), 170.

38. Jacob Böhme, *The Signature of All Things and Other Writings* (Cambridge: James Clarke, 1969), 23.

39. Agamben, *The Signature of All Things*; Singh Uberoi, *The European Modernity*.

40. Foucault, *The Order of Things*, 25.

41. Agamben, *The Signature of All Things*, 36.

42. Eco, *Semiotics and the Philosophy of Language.*

43. Bruno Latour, "When Things Strike Back: A Possible Contribution of 'Science Studies' to the Social Sciences," *British Journal of Sociology* 51, no. 1 (2000): 10–23.

44. Latour, "When Things Strike Back," 115.

45. Mangano, "Understanding the Standard Model," 9.

46. Latour, "When Things Strike Back," 121.

47. Georges Canguilhem, *The Normal and the Pathological* (New York: Zone, 1989).

48. Knorr-Cetina, *Epistemic Cultures*, 146.

49. Knorr-Cetina, *Epistemic Cultures*, 52–53.

50. Evelyn Fox Keller, *A Feeling for the Organism: The Life and Work of Barbara Mcclintock* (New York: W. H. Freeman, 1993).

3. ON ORIENTATION

1. Jorge Luis Borges, *The Aleph and Other Stories* (New York: Penguin, 2004), 105–6.

2. Emile Durkheim, *The Elementary Forms of the Religious Life*, trans. Joseph Ward Swain (1915; repr., New York: Free Press, 1965), 23–24.

3. Martin Gardner, *The New Ambidextrous Universe: Symmetry and Asymmetry from Mirror Reflections to Superstrings* (New York: W. H. Freeman, 1990); and Ernst Cassirer, *The Philosophy of Symbolic Forms*, vol. 2 (New Haven, CT: Yale University Press, 1953).

4. Robert Hertz, "The Pre-eminence of the Right Hand: A Study in Religious Polarity," in Rodney Needham, ed., *Right and Left; Essays on Dual Symbolic Classification* (1909; repr., Chicago: University of Chicago Press, 1973), 3.

5. Hertz, "The Pre-eminence of the Right Hand," 16.

6. Hertz, "The Pre-eminence of the Right Hand," 15.

7. E. E. Evans-Pritchard, "Foreword" in Needham, *Right and Left*, x.

8. Although is possible to use gravity as a cue to define down and up or to use the size of the hydrogen atom to define length, no such basis exists to unambiguously differentiate right and left, or at least until the discovery of parity violation, as will be shown later.

9. Hans Reichenbach, *The Rise of Scientific Philosophy* (Berkeley: University of California Press, 1951), 134.

10. In proposition 6.36111 of the *Tractatus*, Wittgenstein argues that a right-hand glove can indeed be worn on the left hand—simply by turning it inside out. The upshot is that in a Euclidean space of rigid rotations and translations, two hands cannot be superimposed or mapped onto each other. One requires an extra dimension in space—for a rotation in (n + 1) dimension—to make them congruent. This is exactly what a Möbius strip achieves: two non-superimposable objects in a three-dimensional space become identical when one of them is rotated or turned over in a fourth, although physically unattainable, dimension. Ludwig Wittgenstein, *Tractatus Logico-Philosophicus*, trans. David Pears and Brian McGuinness (London: Routledge, 2001), 81.

11. Immanuel Kant, "Concerning the Ultimate Ground of the Differentiation of Directions in Space," in *Theoretical Philosophy, 1755–1770*, trans. David Walford and Ralf Meerbote (1768; repr., Cambridge: Cambridge University Press, 1992); John Earman, *World Enough and Space-Time: Absolute Versus Relational Theories of Space and Time* (Cambridge, MA: MIT Press, 1989); and Katherine Brading and Elena Castellani, *Symmetries in Physics: Philosophical Reflections* (Cambridge: Cambridge University Press, 2003).

12. Mara Beller, *Quantum Dialogue: The Making of a Revolution* (Chicago: University of Chicago Press, 1999); and Karl Popper, "Quantum Mechanics Without 'The Observer,'" in Mario Bunge, *Quantum Theory and Reality* (New York: Springer-Verlag, 1967).

13. Werner Heisenberg, *Physics and Philosophy: The Revolution in Modern Science* (New York: Harper, 1958), 29.

14. Walter Greiner and Berndt Müller, *Gauge Theory of Weak Interactions* (New York: Springer, 2009).

15. Ferdinand de Saussure, *Course in General Linguistics*, trans. Wade Baskin (New York: Columbia University Press, 2011).

16. Ludwig Wittgenstein, *Philosophical Investigations*, trans. G. E. M. Anscombe (Oxford: Blackwell, 2001), section 243.

17. This aspect of the problem is more familiarly known as the "Ozma problem," following Martin Gardner's popular discussion. See Gardner, *The New Ambidextrous Universe*.

18. Richard Phillips Feynman, *The Character of Physical Law* (Cambridge, MA: MIT Press, 1965), 107.

19. Brading and Castellani, *Symmetries in Physics*; Gian Francesco Giudice, *A Zeptospace Odyssey: A Journey Into the Physics of the LHC* (Oxford:

Oxford University Press, 2010); and L. V. Tarasov, *This Amazingly Symmetrical World* (Moscow: Mir, 1986).

20. Eugene Paul Wigner, *Symmetries and Reflections; Scientific Essays of Eugene P. Wigner* (Bloomington: Indiana University Press, 1967).

21. Discrete symmetry refers to noncontinuous transformations, such as mirror reflection. In addition to parity, time and charge conjugation are other discrete symmetries.

22. The weak force is responsible for the disintegration and decay of matter via radioactivity.

23. Allan Franklin, *The Neglect of Experiment* (Cambridge: Cambridge University Press, 1986).

24. T. D. and C. N. Yang Lee, "Question of Parity Conservation in Weak Interactions," *Physical Review* 104 (1956): 254–58; and A. K. Wroblewski, "The Downfall of Parity: The Revolution That Happened Fifty Years Ago," *Acta Physica Polonica* 39 no. 2 (2008) : 251–64.

25. Two other similar experiments were performed around the same time, which confirmed Madame Wu's findings, and brought for T. D. Lee and C. N. Yang the 1957 Nobel Prize in Physics.

26. Earman, *World Enough and Space-Time*; and Gardner, *The New Ambidextrous Universe*.

27. Steven Weinberg, *Dreams of a Final Theory* (New York: Pantheon, 1992), 245.

28. Immanuel Kant, *Critique of Pure Reason*, trans. Paul Guyer and Allen W. Wood (Cambridge: Cambridge University Press, 1998).

29. Kant, "Concerning the Ultimate Ground of the Differentiation of Directions in Space."

30. H. G. Alexander, ed., *The Leibniz-Clarke Correspondence: Together with Extracts from Newton's* Principia *and* Opticks (New York: Philosophical Library, 1956), 25–26.

31. Alexander, *The Leibniz-Clarke Correspondence*, 25–26.

32. Hermann Weyl, *Symmetry* (Princeton, NJ: Princeton University Press, 1952). Note that at the time when Weyl is expressing these sentiments, parity violation, which puts into question the isotropy and homogeneity of space, had not yet been experimentally discovered.

33. Weyl, *Symmetry*, 22.

34. Needham, *Right and Left*.

35. Alexandre Koyre, *From the Closed World to the Infinite Universe* (Baltimore: Johns Hopkins Press, 1957); and Frances Amelia Yates, *Giordano*

Bruno and the Hermetic Tradition (Chicago: University of Chicago Press, 1964).

36. Renè Descartes, *Principles of Philosophy* (Lewiston, NY: E. Mellen, 1988), 19.

37. R. Ariew and D. Garber, eds., *G. W. Leibniz: Philosophical Essays* (Indianapolis: Hackett, 1989).

38. Gottlob Frege, *The Foundations of Arithmetic; A Logico-Mathematical Enquiry Into the Concept of Number* (New York: Harper, 1960); and Bertrand Russell, *Our Knowledge of the External World as a Field for Scientific Method in Philosophy* (London: Allen & Unwin, 1926). The scholastic frame of subject-predicate analysis (i.e., every proposition is about a "substance" with definite "attributes") makes it impossible to admit the "reality" of relational attributes like to the left of, the father of, and so on.

39. John Locke, *An Essay Concerning Human Understanding* (Cleveland, OH: Meridian, 1964), 202.

40. Hertz, "The Pre-eminence of the Right Hand," 21–22. The rational part follows Kant's argument that we are all endowed with the sense of difference of our body's two sides, right and left.

41. Hertz, "The Pre-eminence of the Right Hand," 22.

42. Hertz, "The Pre-eminence of the Right Hand," 20

43. Hertz, "The Pre-eminence of the Right Hand," 10.

44. Hertz, "The Pre-eminence of the Right Hand," 10.

45. Hertz, "The Pre-eminence of the Right Hand," 10.

46. Incidentally, it is the mathematical formalism of Dirac's equation—the gamma 5 matrices—which gives birth to the concept of chirality in particle physics. Weyl observed that the four-component Dirac equation splits into two independent equations by taking the zero-mass limit of the electron, which he termed, "the left-handed" and "right-handed" spinors. There was no impetus for splitting the wave function into two different components from any experiment; calling them left or right also did not signify anything at the time. It was motivated purely by the algebra of matrices. In the process of connecting the chiral components with spinors and the deep structure of space-time, Weyl raises the important question of mass in relativistic quantum physics. See Hermann Weyl, "Gravitation and the Electron," *Proceedings of the National Academy of Sciences of the United States of America* 15, no. 4 (1929): 323–34.

47. Andrei Sakharov was the first physicist to propose CP violation as one of the three formal conditions for "baryogenesis," or the process of creation of matter in the early universe.

48. Hertz, "The Pre-eminence of the Right Hand," 13.

49. See Mary Douglas, *Implicit Meaning: Selected Essays in Anthropology* (1975; repr., London: Routledge, 1999), for a comprehensive discussion of the "social determinants" of cosmology.

50. Ludwig Wittgenstein et al., *Remarks on the Foundations of Mathematics* (Cambridge, MA: MIT Press, 1967), section 74.

51. The evidence of parity violation in electroweak interactions shows us, in Wittgenstein's words, how deep are human conventions. Another pithy aphorism of his conveys the same import: "to a necessity in the world," he writes, "there corresponds an arbitrary rule in language." Desmond Lee, ed., *Wittgenstein's Lectures, Cambridge, 1930–1932: From the Notes of John King and Desmond Lee* (Oxford: Blackwell, 1980), 57.

4. THE CYCLE OF WORK

1. This chapter was previously published in a slightly different form as "Science and the Large Hadron Collider: A Probe Into Instrumentation, Periodization and Classification," *Dialectical Anthropology* 36, nos. 3–4 (2012): 291–316. I am grateful to the editors and publishers of the journal for allowing me to make use of the article here.

2. The LHC circular accelerator is not a perfect circle, but rather it is split into eight distinct parts—or sectors—composed of eight arcs and eight straight sections. The sectors are the working units of the LHC: magnet installation, hardware commissioning, and powering, which all take place sector by sector.

3. Joseph Masco, "Lie Detectors: On Secrets and Hypersecurity in Los Alamos," *Public Culture* 14, no. 3 (2002): 451.

4. Victor Turner, "Social Dramas and Stories About Them," *Critical Inquiry* 7, no. 1 (1980): 149.

5. Robert Aymar, email to all CERN personnel, subject line: Report on 19th Sept 2008 incident at LHC—rapport sur l'incident du 19 Sept 2008 au LHC.

6. Mike Lamont is the leader of the Operations Group (Beams Department), which is responsible for the technical infrastructure and operation

of all CERN present and future accelerators. On April 23, 2010, Lamont presented the "LHC status report" at CERN's main auditorium.

7. The BBC ran an interesting article on February 17, 2009, entitled, "Race for God Particle Heats Up," singling out the prospects of the Tevatron.

8. Karl Marx and Friedrich Engels, *Capital: A Critique of Political Economy* (New York: International, 1967).

9. Oliver Bruning et al., ed., *LHC Design Report* (Geneva: CERN, 2004)

10. Articles appearing in various issues of the *CERN Courier*, the monthly journal brought out by CERN, provides a wonderful account of the different phases of the LHC design and development process, such as "French Green Light for LHC Civil Engineering" (October 1998), "Model Magnet for CERN's LHC Reaches 250 T/m in Japan" (April 1999), "First Test Beams Are Delivered for the LHC" (September 2000), "Going Into the Cold: LHC Systems Reach an Important Milestone" (December 2001), "High-Energy Accelerators Look to R&D" (June 2001), "Last LHC Magnets from Siberia Reach CERN" (September 2001), and "CERN Reacts to Increased LHC Costs" (January/February 2002). I have relied heavily on issues appearing between 1994 and 2007 of the *CERN Courier* for much of the history of the LHC.

11. LHC Study Group, *The Large Hadron Collider, Conceptual Design* (CERN/AC/95-05: 1995), 3.

12. See Alexander W. Chao and Weiren Chou, ed., *Reviews of Accelerator Science and Technology* (Singapore: World Scientific, 2008); Stanley M. Livingston, *Particle Accelerators: A Brief History* (Cambridge, MA: Harvard University Press, 1969); Enrico Persico, Ezio Ferrari, and Sergio E. Segre, *Principles of Particle Accelerators* (New York: W. A. Benjamin, 1968); and A. M. Sessler and E. J. N. Wilson, *Engines of Discovery: A Century of Particle Accelerators* (Hackensack, NJ: World Scientific, 2007).

13. The types of particles used for collisions, whether leptons or hadrons, also play a role in the alternating and complementary logic of accelerator development. Hadron colliders, using protons or neutrons, are especially suited for discoveries as they allow for a certain breadth in understanding the inner constitution of matter. Lepton machines, which use electrons or positrons, are more appropriate for precision measurements of particles after their discovery.

14. Lyndon Evans, *The Large Hadron Collider: A Marvel of Technology* (Lausanne: EPFL Press, 2009), 74. The construction, transportation,

and installation of dipole magnets, which had to cross a number of logistical, civil engineering, legal, and financial hurdles, is a fascinating story in itself. Also see the following articles from *CERN Courier*, "LHC Dipole Production Begins to Take Off" (January 2004), "LHC Dipole Installation Gets to Half-Way Mark" (September 2006), "The Longest Journey: The LHC Dipoles Arrive on Time" (October 2006), and "The Last Dipole Makes Its Descent" (June 2007).

15. Evans, *The Large Hadron Collider*, 74.

16. CERN had decided early on that the LHC was to be constructed in the LEP tunnel for optimum use of existing infrastructure and keeping costs down. This also imposed a number of constraints on the LHC machine design and layout. For more detail, see Evans, *The Large Hadron Collider*.

17. J. P. S. Uberoi brought to my attention the critical significance of this moment of crisis, which showed the splitting up of laboratory work cycle into discrete operations. It is his insight on how time gives form to the distribution of work, and its singular importance to anthropological fieldwork, which leaves a trace in this chapter.

18. Louis Althusser and Etienne Balibar, *Reading* Capital (London: NLB, 1970).

19. Althusser and Balibar, *Reading* Capital, 229.

20. "Commissioning of the LHC Magnet Powering System in 2009," *Proceedings of IPAC'10*, Kyoto, Japan.

21. CERN Annual Report, 2009.

22. Katie Yurkewicz, "CERN's New LHC Plan: Two Years at 3.5 TeV," *Symmetry* 7, no. 5 (2010).

23. From Steve Myers' presentation on February 24, 2009, at CERN summarizing the Chamonix Workshop.

24. CERN Annual Report, 2008, 25–26.

25. Gian Francesco Giudice, *A Zeptospace Odyssey: A Journey Into the Physics of the LHC* (Oxford: Oxford University Press, 2010), 94.

26. Luminosity can be computed from the number of particles circulating in the two beam directions by taking their product $N_1.N_2$ and dividing it by the revolution frequency and the transverse section of the beam. One obtains a number whose units are inverse area (the beam size) times inverse time in seconds (the frequency), which is usually expressed in cgs units, $cm^{-2} s^{-1}$. In his popular blog online, *A Quantum Diaries Survivor*, Italian physicist, Tomaaso Dorigo, submits that "luminosity is not

just a number with which machinists boast about their gadget. With it, you can actually compute the rate of production of any given process, if you know its cross section." Dorigo, "Multiple Interactions at LHC: An Exercise in Elementary Statistics," *A Quantum Diaries Survivor* (blog), February 13, 2008, https://dorigo.wordpress.com/2008/02/13/multiple -interactions-at-lhc-an-exercise-in-elementary-statistics/.

27. As the ATLAS-LBNL group leader, Ian Hinchliffe, along with Mur-dock (Gil) Gilchriese, approved my affiliation at CERN, which enabled me to maintain a connection with the European laboratory for up to seven years.

28. Posted by Ted Kolberg, February 21, 2009, at http://blogs.uslhc.us/lhc -luminosity-and-energy.

29. "People: An International Spirit," CERN, http://public.web.cern.ch/public /en/People/Experimentalists-en.html.

30. Karl Marx et al., *The German Ideology: Parts I and III* (New York: International Publishers, 1947), 20.

31. Jurgen Habermas, *Theory and Practice* (Boston: Beacon Press, 1973), 9.

32. Paul Rabinow, *Making PCR: A Story of Biotechnology* (Chicago: University of Chicago Press, 1996); and Robert Kohler, *Landscapes and Lab-scapes: Exploring the Lab-Field Border in Biology* (Chicago: University of Chicago Press, 2002).

33. Steven Shapin, "Science and the Modern World," in *The Handbook of Science and Technology Studies*, ed. Edward J. Hackett (Cambridge, MA: MIT Press, 2008), 443.

34. Sherry B. Ortner, "Theory in Anthropology Since the Sixties," *Comparative Studies in Society and History* 26, no. 1 (1984): 126–66; and Theodore R. Schatzki, Karin Knorr-Cetina, and Eike von Savigny, eds., *The Practice Turn in Contemporary Theory* (London: Routledge, 2001).

35. Ian Hacking, *Representing and Intervening: Introductory Topics in the Philosophy of Natural Science* (Cambridge: Cambridge University Press, 1983).

36. Peter Galison, *Image and Logic: A Material Culture of Microphysics* (Chicago: University of Chicago Press, 1997), 783.

37. John Duprè, *The Disorder of Things: Metaphysical Foundations of the Disunity of Science* (Cambridge, MA: Harvard University Press, 1993); Peter Galison and David J. Stump, *The Disunity of Science: Boundaries, Contexts, and Power* (Stanford, CA: Stanford University Press, 1996); and Hacking, *Representing and Intervening*.

5. ART, SCIENCE, AND POSTMODERNISM

1. Alan Sokal, "A Physicist Experiments with Cultural Studies," *Lingua Franca* (May/June 1996): 62–64.

2. Sokal's parody article has since been followed by numerous books questioning the malaise afflicting critical theory, postmodernism, and science and technology studies. See also Alan Sokal and Jean Bricmont, *Fashionable Nonsense: Postmodern Intellectuals' Abuse of Science* (New York: Picador, 1998); and Alan Sokal, *Beyond the Hoax: Science, Philosophy, and Culture* (Oxford: Oxford University Press, 2008).

3. Paul Boghossian, *Fear of Knowledge: Against Relativism and Constructivism* (Oxford: Clarendon Press, 2006).

4. Sokal and Bricmont, *Fashionable Nonsense*.

5. Leonard Shlain, *Art and Physics: Parallel Visions in Space, Time, and Light* (New York: Morrow, 1991); and Arthur I. Miller, *Insights of Genius: Imagery and Creativity in Science and Art* (Cambridge, MA: MIT Press, 2000).

6. Katarina Anthony, "Cern Has a New Cultural Policy," *The Bulletin*, December, 13, 2010.

7. Ariane Koek, "Viewpoint: Collide—A Cultural Revolution," *CERN Courier*, July 7, 2010.

8. Jenny Marder, "At Cern, Art Collides with Science," PBS, http://www.pbs.org/newshour/updates/science/july-dec10/cern_12-01.html.

9. "Where Science and Art Collide," *The Bulletin*, September 7 and 14, 2009, 4.

10. Christine Sutton, "Creativity Across Cultures," *CERN Courier* 58, no. 6 (July/August 2018).

11. Michelangelo Mangano, "Visible and Invisible in Modern Physics" (CERN, Theoretical Physics Department, Arts and Science Across Italy, September 9–13, 2018), https://indico.cern.ch/event/729517/contributions/3071047/attachments/1714957/2766287/Mangano-ASinItaly.pdf

12. For more details on the analogy, see Koek's online blog. Ariane Koeck, "It's Only a Matter of Time . . . ," *Beauty Quark* (blog), August 19, 2010, in http://wwwbeautyquark-beautyquark.blogspot.com/.

13. John Ruskin, *Modern Painters*, vol. 5 (London: Dent, 1920–1935).

14. Ariane Koek, "In/visible: The Inside Story of the Making of Arts at CERN," *Interdisciplinary Science Reviews: Art and Science* 42, no. 4 (2017): 1.

15. Arthur Coleman Danto, *Beyond the Brillo Box: The Visual Arts in Post-Historical Perspective* (New York: Farrar Straus Giroux, 1992); and George Santayana, *The Life of Reason; or, the Phases of Human Progress* (New York: Scribner's, 1905).

16. See Charles Taylor, *Philosophy and the Human Sciences* (Cambridge: Cambridge University Press, 1985); and for an adumbration of "The Practice Turn," see Sherry Ortner, "Theory in Anthropology Since the Sixties," *Comparative Studies in Society and History* 26, no. 1 (1984): 126–66.

17. See, for instance, Herbert Marcuse, *One Dimensional Man: Studies in the Ideology of Advanced Industrial Society* (Boston: Beacon Press, 1964); Jacques Ranciere, *Aesthetics and Its Discontents* (Cambridge: Polity, 2009); and Theodore W. Adorno, *Aesthetics* (Cambridge: Polity, 2018) on the value of the critical function of art and aesthetics.

18. Bert Ulrich, "NASA and the Arts," NASA, https://www.nasa.gov/50th/50th_magazine/arts.html.

19. For illustrations of the art works, see Piers Bizony, *The Art of NASA: The Illustrations That Sold the Missions* (Beverly, MA: Motorbooks, 2020).

20. Keith Cowing, "NASA's First and Last Artist in Residence?," NASA Watch, June 21, 2005, http://nasawatch.com/archives/2005/06/nasas-first-and-last-artist-in-residence.html.

21. Melissa Ragain, "From Organization to Network: MIT's Center for Advanced Visual Studies," *X-tra* 3 (Spring 2012), http://x-traonline.org/article/from-organization-to-network-mits-center-for-advanced-visual-studies.

22. Ragain, "From Organization to Network."

23. Ragain, "From Organization to Network." In 2009, the CAVS was merged with the Visual Arts Program and renamed as Program in Art, Culture, and Technology. For more details of the CAVS blueprint on the nexus of art and science with reference to MIT's military work, see also John Beck and Ryan Bishop, ed., *Cold War Legacies: Systems, Theory, Aesthetics* (Edinburgh: Edinburgh University Press, 2016); Anne Collins Goodyear, "György Kepes, Billy Klüver, and American Art of the 1960s: Defining Attitudes Toward Science and Technology," *Science in Context* 17, no. 4 (2004): 611–35; John R. Blakinger, "The Aesthetics of Collaboration: Complicity and Conversion at MIT's Center for Advanced Visual Studies," Tate Papers no. 25 (2016), https://www.tate.org.uk/research/publications/tate-papers/25/aesthetics-of-collaboration.

24. It is a familiar sight to find promotional material at CERN, such as posters and photos, featuring chalkboards dotted with illegible formulae, with a pair of scientists standing facing it, with one of them scribbling while the other is looking at it in deep thought. For a fine description of the typical inner spaces of a high-energy physics laboratory, see also chapter 1 of Sharon Traweek's *Beamtimes and Lifetimes: The World of High Energy Physicists* (Cambridge, MA: Harvard University Press, 1988).

25. Lorraine Walsh, "Art and Outreach," *SCGP News*, Special Issue XV (2020): 70–72.

26. Walsh, "Art and Outreach," 67

27. Ragain, "From Organization to Network."

28. These "glass box" lunches—so named because of the portion of the main cafeteria marked off by a glass wall—are often organized when political dignitaries, famous scientists, and other luminaries visit CERN for book launches and talks.

29. Steven Shapin, *The Scientific Life: A Moral History of a Late Modern Vocation* (Chicago: University of Chicago Press, 2009). I refer to this remarkable work also in the context of the shift in scientific life from a "calling" to a job.

30. Clare Wiley, "Dancing in the CERN Library—Let's Get Physical," *The Guardian*, July 16, 2012, https://www.theguardian.com/stage/2012/jul/16/cern-dance-strangels-sciart.

31. Sutton, "Creativity Across Cultures."

32. Rolf Heuer, "Foundations for the Future," *The Bulletin*, December 13, 2010.

33. Anthony, "Cern Has a New Cultural Policy."

34. Massimo Giovannini is a theoretical physicist from the National Institute for Nuclear Physics, in Milan, and was partially based at CERN.

35. Gian Francesco Giudice, *A Zeptospace Odyssey: A Journey Into the Physics of the LHC* (Oxford: Oxford University Press, 2010), 146.

36. P. A. M. Dirac, "The Relation Between Mathematics and Physics," in *The Collected Works of P. A. M. Dirac, 1924–1948*, ed. R. H. Dalitz (Cambridge: Cambridge University Press, 1995); and Frank Wilczek, *The Lightness of Being: Mass, Ether, and the Unification of Forces* (New York: Basic Books, 2008).

37. Giudice, *A Zeptospace Odyssey*, 147.

38. David J. Gross, "The Role of Symmetry in Fundamental Physics," *Proceedings of the National Academy of Sciences of the United States of America* 93, no. 25 (1996): 14256–59; Eugene P. Wigner, *Symmetries and Reflections: Scientific Essays of Eugene P. Wigner* (Bloomington: Indiana University Press, 1967); and Steven Weinberg, *Dreams of a Final Theory* (New York: Pantheon, 1992).

39. Ian Stewart, *Why Beauty Is Truth: A History of Symmetry* (New York: Basic Books, 2007), 277.

40. Markus Nordberg is a physicist who is also the resources coordinator affiliated with the ATLAS experiment at CERN.

41. Thomas S. Kuhn, *The Essential Tension: Selected Studies in Scientific Tradition and Change* (Chicago: Chicago University Press 1977), 342.

42. Immanuel Kant, *The Critique of Judgement*, trans. James Creed Meredith (Oxford: Clarendon Press, 1952).

43. Sutton, "Creativity Across Cultures."

44. Zygmunt Bauman, *Intimations of Postmodernity* (London: Routledge, 1992).

45. Todd Gitlin, "Postmodernism: Roots and Politics," in *Cultural Politics in Contemporary America*, ed. Ian Angus and Sut Jhally (London: Routledge, 1989), 351.

46. Joseph Natoli and Linda Hutcheon, ed., *A Postmodern Reader* (Albany: State University of New York Press, 1993); Bruno Latour, *We Have Never Been Modern* (Cambridge, MA: Harvard University Press, 1993); and J. P. S. Uberoi, *The European Modernity: Science, Truth and Method* (Delhi: Oxford University Press, 2002).

47. Wiley, "Dancing in the CERN Library."

48. Hailey Reissman, "Indie Band Deerhoof Experiment with Sound at CERN," CERN, September 18, 2015, https://home.cern/news/news/cern /indie-band-deerhoof-experiment-sound-cern; and *CERN Courier* 58, no. 6 (July/August 2018), https://cds.cern.ch/record/2628313/files /CERNCourier2018JulAug-digitaledition.pdf.

49. Sokal and Bricmont, *Fashionable Nonsense*.

50. John Ellis, "Answering Gauguin's Questions with the LHC," CERN, https://indico.cern.ch/event/145296/contributions/1381149/attachments /136914/194263/KNUST.pdf.

EPILOGUE

1. Ludwig Wittgenstein, *Tractatus Logico-Philosophicus* (1961; repr., London: Routledge, 2001).

2. See J. P. S. Uberoi, *The Other Mind of Europe: Goethe as a Scientist* (Delhi: Oxford University Press, 1984); and Bernard d'Espagnat, *On Physics and Philosophy* (Princeton, NJ: Princeton University Press, 2006) for more details on the source of competing claims between optics (primary qualities) and colors (secondary qualities) over the rainbow.

3. Michel Foucault, "Two Lectures," in *Michel Foucault Power/Knowledge: Selected Interviews and Other Writings 1972–1977*, ed. C. Gordon (New York: Pantheon, 1980), 81. According to Deleuze and Guattari, everywhere among the sciences are two tendencies, a "nomad science," which is nonlinear, works from the ground up, and is decentered and unstable, and a "royal science," which is driven by rational, axiomatic procedures and reflective of the political mainstream. My point is that we are obliged to counter the elevation of objectivity by not confining ourselves to nomad science. Gilles Deleuze and Felix Guattari, *A Thousand Plateaus: Capitalism and Schizophrenia* (Minneapolis: University of Minnesota Press, 1987).

4. Michel Foucault, *The Order of Things: An Archaeology of the Human Sciences* (New York: Vintage, 1970), 29.

5. Giorgio Agamben, *The Signature of All Things: On Method* (New York: Zone, 2009), 43.

6. J. P. S. Uberoi, *Science and Culture* (Delhi: Oxford University Press, 1978), 43.

7. See Daniel Little, *A New Social Ontology of Government* (Cham: Palgrave, 2020) for a contemporary dive into science policy and government action.

8. The standpoint of relations, and how removed they are from empirical occurrence, is illustratively presented in Russell's example of two propositions "there is cheese" and "there is no cheese." The two propositions appear to share the same form in that both refer to the real world. However, no empirical experience corresponds to the second proposition. Negation, or contradiction, and broadly speaking, the sphere of logical relations, can never be encountered in the world of things. See Bertrand Russell, *An Inquiry Into Meaning and Truth* (London: Routledge, 1992), 73–74.

9. Claude Lévi-Strauss, *Structural Anthropology* (New York: Basic Books, 1963); Philippe Descola, *Beyond Nature and Culture* (Chicago: University of Chicago Press, 2013); and Emile Durkheim, *The Elementary Forms of the Religious Life* (New York: Free Press, 1965), 21.

10. Mary Douglas, *Implicit Meanings: Essays in Anthropology* (1975; repr., London: Routledge & Kegan, 1999). We cannot neglect to recall Traweek's pioneering work putting out a Dukheimian approach with a study of the cosmology of high-energy physics. Sharon Traweek, *Beamtimes and Lifetimes: The World of High-Energy Physicists* (Cambridge, MA: Harvard University Press, 1988).

REFERENCES

Agamben, Giorgio. *The Signature of All Things: On Method*. Trans. Luca di Santo and Kevin Attell. New York: Zone, 2009.

Adorno, Theodore W. *Aesthetics*. Cambridge: Polity, 2018.

Alexander, H. G., ed. *The Leibniz-Clarke Correspondence: Together with Extracts from Newton's* Principia *and* Opticks. New York: Philosophical Library, 1956.

Althusser, Louis, and Etienne Balibar. *Reading* Capital. London: NLB, 1970.

Anthony, Katarina. "Cern Has a New Cultural Policy." *The Bulletin*, nos. 50 and 51, December 13, 2010, https://cds.cern.ch/record/1312210?ln=en.

ATLAS. "Technical Proposal." 1994.

ATLAS Collaboration. "Luminosity Determination in Pp Collisions at S = 7 Tev Using the Atlas Detector at the LHC." *arXiv:1101.2185v1 [hep-ex]* (2011).

Ariew, Roger, and D. Garber, eds. *G. W. Leibniz: Philosophical Essays*. Indianapolis: Hackett, 1989.

Bachelard, Gaston. *The New Scientific Spirit*. Boston: Beacon, 1984.

Barad, Karen Michelle. *Meeting the Universe Halfway: Quantum Physics and the Entanglement of Matter and Meaning*. Durham, NC: Duke University Press, 2006.

Bauman, Zygmunt. *Culture as Praxis*. London: Routledge and Kegan, 1973.

——. *Intimations of Postmodernity*. London: Routledge, 1992.

Beck, John, and Ryan Bishop, eds. *Cold War Legacies: Systems, Theory, Aesthetics*. Edinburgh: Edinburgh University Press, 2016.

Beller, Mara. *Quantum Dialogue: The Making of a Revolution*. Chicago: University of Chicago Press, 1999.

Bennett, G., and P. Rabinow. "Invitation: Synthetic Biology and Human Practices: A Problem." 2008. http://cnx.org/content/m18812/latest/.

Benveniste, Émile. *Problems in General Linguistics.* Coral Gables, FL: University of Miami Press, 1971.

Biagioli, Mario, ed. *The Science Studies Reader.* New York: Routledge, 1999.

Bizony, Piers. *The Art of NASA: The Illustrations That Sold the Missions.* Beverly, MA: Motorbooks, 2020.

Bjorken, J. D. "The Future and Its Alternatives." In www.slac.stanford .edu/. . ./BJ_The%20Future%20and%20Its%20Alternatives.pdf.

Blakinger, John R. "The Aesthetics of Collaboration: Complicity and Conversion at MIT's Center for Advanced Visual Studies." *Tate Papers* no. 25 (2016). https://www.tate.org.uk/research/tate-papers/25/aesthetics-of -collaboration.

Bloor, D. "Anti-Latour." *Studies in History and Philosophy of Science* 30, no. 1 (1999): 81–112.

——. "Durkheim and Mauss Revisited: Classification and the Sociology of Knowledge." *Studies in History and Philosophy of Science* 13, no. 4 (1982): 267–97.

Boghossian, Paul. *Fear of Knowledge: Against Relativism and Constructivism.* Oxford: Clarendon, 2006.

Böhme, Jacob. *The Signature of All Things and Other Writings.* Cambridge: James Clarke, 1969.

Borges, Jorge Luis. *The Aleph and Other Stories.* New York: Penguin, 2004.

Bourdieu, Pierre. *Outline of a Theory of Practice.* Cambridge: Cambridge University Press, 1977.

Brading, Katherine, and Elena Castellani. *Symmetries in Physics: Philosophical Reflections.* Cambridge: Cambridge University Press, 2003.

Brumfiel, Geoff. "LHC Students Face Data Drought." *Nature* 460, no. 7255 (2009): 558.

Bruning, Oliver, and Paul Collier. "Building a Behemoth." *Nature* 448, no. 7151 (2007): 285.

Bruning, Oliver, et al., ed. *LHC Design Report.* Geneva: CERN, 2004.

Bunge, Mario. *Quantum Theory and Reality.* New York: Springer-Verlag, 1967.

Canguilhem, Georges. *The Normal and the Pathological.* New York: Zone, 1989.

——. *A Vital Rationalist: Selected Writings from Georges Canguilhem.* New York: Zone, 2000.

Chao, Alexander W., and Weiren Chou, eds. *Reviews of Accelerator Science and Technology*. Singapore: World Scientific, 2008.

Cassirer, Ernst. *The Philosophy of Symbolic Forms*. New Haven, CT: Yale University Press, 1953.

CERN. "Where Science and Art Collide." *The Bulletin*, nos. 37–38, September 7, 2009. https://cdsweb.cern.ch/record/1204806.

Collins, H. M. *Gravity's Shadow: The Search for Gravitational Waves*. Chicago: University of Chicago Press, 2004.

——. *Are We All Scientific Experts Now?* Cambridge: Polity, 2014.

Dalton, Russell J. *Critical Masses: Citizens, Nuclear Weapons Production, and Environmental Destruction in the United States and Russia*. Cambridge, MA: MIT Press, 1999.

Danto, Arthur Coleman. *Beyond the Brillo Box: The Visual Arts in Post-Historical Perspective*. New York: Farrar Straus Giroux, 1992.

Deleuze, Gilles, and Felix Guattari. *A Thousand Plateaus: Capitalism and Schizophrenia*. Minneapolis: University of Minnesota Press, 1987.

Derrida, Jacques. *Margins of Philosophy*. Chicago: University of Chicago Press, 1982.

de Saussure, Ferdinand. *Course in General Linguistics*. Trans. Wade Baskin. New York: Columbia University Press, 2011.

Descartes, Renè. *Principles of Philosophy*. Lewiston, NY: E. Mellen, 1988.

Descola, Philippe. *Beyond Nature and Culture*. Chicago: University of Chicago Press, 2013.

Dirac, P. A. M., *The Collected Works of P. A. M. Dirac, 1924–1948*. Ed. R. H. Dalitz. Cambridge: Cambridge University Press, 1995.

Dorigo, Tommaso. "Particle Physics in 2020." *A Quantum Diaries Survivor: Private Thoughts of a Physicist and Chessplayer* (blog). *Science 2.0*, January 18, 2010. https://www.science20.com/quantum_diaries_survivor/particle _physics_2020.

——. "Multiple Interactions at LHC: An Exercise in Elementary Statistics," *A Quantum Diaries Survivor: Private Thoughts of a Physicist and Chessplayer* (blog), February 13, 2008, https://dorigo.wordpress.com/2008/02/13 /multiple-interactions-at-lhc-an-exercise-in-elementary-statistics/.

Douglas, Mary. *Implicit Meanings: Selected Essays in Anthropology*. 1975. Reprint, London: Routledge, 1999.

Duprè, John. *The Disorder of Things: Metaphysical Foundations of the Disunity of Science*. Cambridge, MA: Harvard University Press, 1993.

Durkheim, Emile. *The Elementary Forms of the Religious Life*. 1915. Reprint, New York: Free Press, 1965.

Durkheim, Emile, and Marcel Mauss. *Primitive Classification*. Chicago: University of Chicago Press, 1963.

Earman, John. *World Enough and Space-Time: Absolute Versus Relational Theories of Space and Time*. Cambridge, MA: MIT Press, 1989.

Eco, Umberto. *A Theory of Semiotics*. Bloomington: Indiana University Press, 1976.

——. *Semiotics and the Philosophy of Language*. Bloomington: Indiana University Press, 1984.

Einstein, Albert. "Geometry and Experience." In *An Expanded Form of an Address to the Prussian Academy of Sciences in Berlin*, January 27, 1921. https://mathshistory.st-andrews.ac.uk/Extras/Einstein_geometry.

Ellis, John. "Beyond the Standard Model with the LHC." *Nature* 448, no. 7151 (2007): 297–301.

d'Espagnat, Bernard. *On Physics and Philosophy*. Princeton, NJ: Princeton University Press, 2006.

European Organization for Nuclear Research. "Introducing the Cern Courier." *CERN Courier* 1, no. 1 (1959): 1.

Evans, Lyndon R. *The Large Hadron Collider: A Marvel of Technology*. Lausanne: EPFL Press, 2009.

Feldman, Burton. *The Nobel Prize: A History of Genius, Controversy, and Prestige*. New York: Arcade, 2000.

Feynman, Richard Phillips. *The Character of Physical Law*. Cambridge, MA: MIT Press, 1965.

Fischer, Michael M. J. *Emergent Forms of Life and the Anthropological Voice*. Durham, NC: Duke University Press, 2003.

Foucault, Michel. *The Order of Things: An Archaeology of the Human Sciences*. New York: Vintage. 1970.

——. "Two Lectures." In *Michel Foucault Power/Knowledge: Selected Interviews and Other Writings 1972–1977*, ed. C. Gordon. New York: Pantheon, 1980.

Franklin, Allan. *The Neglect of Experiment*. Cambridge: Cambridge University Press, 1986.

Franklin, Sarah. "Re-thinking Nature–Culture: Anthropology and the New Genetics." *Anthropological Theory* 3, no. 1 (2003): 65–85.

Frege, Gottlob. *The Foundations of Arithmetic; A Logico-Mathematical Enquiry Into the Concept of Number*. New York: Harper, 1960.

Galison, Peter. *Image and Logic: A Material Culture of Microphysics.* Chicago: University of Chicago Press, 1997.

——. "Ten Problems in History and Philosophy of Science." *ISIS* 99, no. 1 (2008): 111–24.

Galison, Peter, and Bruce William Hevly. *Big Science: The Growth of Large-Scale Research.* Stanford, CA: Stanford University Press, 1992.

Galison, Peter, and David J. Stump. *The Disunity of Science: Boundaries, Contexts, and Power.* Stanford, CA: Stanford University Press, 1996.

Gardner, Martin. *The New Ambidextrous Universe: Symmetry and Asymmetry from Mirror Reflections to Superstrings.* New York: W. H. Freeman, 1990.

Gitlin, Todd. "Postmodernism: Roots and Politics." In *Cultural Politics in Contemporary America,* ed. Ian Angus and Sut Jhally, 347–60. London: Routledge, 1989.

Giudice, Gian Francesco. *A Zeptospace Odyssey: A Journey Into the Physics of the LHC.* Oxford: Oxford University Press, 2010.

Goodyear, Anne Collins. "György Kepes, Billy Klüver, and American Art of the 1960s: Defining Attitudes Toward Science and Technology." *Science in Context* 17, no. 4 (2004): 611–35.

Goethe, Johann Wolfgang von. *Goethe's Botanical Writings.* Honolulu: University of Hawaii Press. 1952.

Greiner, Walter, and Berndt Müller. *Gauge Theory of Weak Interactions.* New York: Springer, 2009.

Gross, David J. "The Role of Symmetry in Fundamental Physics." *Proceedings of the National Academy of Sciences of the United States of America* 93, no. 25 (1996): 14256–59.

Gusterson, Hugh. *Nuclear Rites: A Weapons Laboratory at the End of the Cold War.* Berkeley: University of California Press, 1996.

Gusterson, Hugh. *People of the Bomb: Portraits of America's Nuclear Complex.* Minneapolis: University of Minnesota Press, 2004.

Habermas, Jurgen. *Theory and Practice.* Boston: Beacon, 1973.

Hackett, Edward J. *The Handbook of Science and Technology Studies.* Cambridge, MA: MIT Press, 2008.

Hacking, Ian. *Representing and Intervening: Introductory Topics in the Philosophy of Natural Science.* Cambridge: Cambridge University Press, 1983.

——. *The Social Construction of What?* Cambridge, MA: Harvard University Press, 1999.

Haraway, Donna Jeanne. *Modestwitness@Secondmillennium.Femalemanmeet-soncomouse : Feminism and Technoscience*. New York: Routledge, 1997.

——. *Simians, Cyborgs, and Women: The Reinvention of Nature*. New York: Routledge, 1991.

Heisenberg, Werner. *Physics and Philosophy: The Revolution in Modern Science*. New York: Harper, 1958.

Heuer, Rolf. "Foundations for the Future." *The Bulletin*, nos. 50 and 51, December 13, 2010. https: http://cds.cern.ch/record/1312979.

Holton, Gerald James. *Thematic Origins of Scientific Thought; Kepler to Einstein*. Cambridge, MA: Harvard University Press, 1973.

Ingold, Tim. *Anthropology and/as Education*. Abington: Routledge, 2018.

Jasanoff, Sheila, Gerald E. Markle, James C. Petersen, and Trevor Pinch, eds. *Handbook of Science and Technology Studies*. Thousand Oaks, CA: Sage, 1995.

Kane, G. L. *Supersymmetry: Unveiling the Ultimate Laws of Nature*. Cambridge, MA: Perseus, 2000.

Kant, Immanuel. *The Critique of Judgement*. Trans. James Creed Meredith. Oxford: Clarendon, 1952.

——. *Critique of Pure Reason*. Trans. Paul Guyer and Allen W. Wood. Cambridge: Cambridge University Press, 1998.

——. *Theoretical Philosophy, 1755–1770*. Trans. David Walford and Ralf Meerbote. Cambridge: Cambridge University Press, 1992.

Keller, Evelyn Fox. *A Feeling for the Organism: The Life and Work of Barbara Mcclintock*. New York: W. H. Freeman, 1993.

Kestenbaum, David. "What Is Electroweak Symmetry Breaking, Anyway?" *FermiNews* 21, no. 2 (1998): 1–3, 8, 11.

Kirksey, S. E., and S. Helmreich. "The Emergence of Multispecies Ethnography." *Cultural Anthropology* 25, no. 4 (2010): 545–76.

Knorr-Cetina, K. *Epistemic Cultures: How the Sciences Make Knowledge*. Cambridge, MA: Harvard University Press, 1999.

——. *The Manufacture of Knowledge: An Essay on the Constructivist and Contextual Nature of Science*. Oxford: Pergamon, 1981.

Koeck, Ariane. "It's Only a Matter of Time . . ." *Beauty Quark* (blog), August 19, 2010. http://wwwbeautyquark-beautyquark.blogspot.com.

——. "Viewpoint: Collide—a Cultural Revolution." Opinion, *CERN Courier*, July 7, 2010.

Kohler, Robert. *Landscapes and Labscapes: Exploring the Lab–Field Border in Biology*. Chicago: University of Chicago Press, 2002.

Koyre, Alexandre. *From the Closed World to the Infinite Universe*. Baltimore: Johns Hopkins Press, 1957.

Kuhn, Thomas S. *The Essential Tension: Selected Studies in Scientific Tradition and Change*. Chicago: University of Chicago Press, 1977.

——. *The Structure of Scientific Revolutions*. Chicago: University of Chicago Press, 1996.

Latour, B. "For David Bloor . . . and Beyond: A Reply to David Bloor's Anti-Latour." *Studies in History and Philosophy of Science* 30, no. 1 (1999): 113–29.

Latour, Bruno. *We Have Never Been Modern*. Cambridge, MA: Harvard University Press, 1993.

——. "When Things Strike Back: A Possible Contribution of 'Science Studies' to the Social Sciences." *British Journal of Sociology* 51, no. 1 (2000): 107–23.

Latour, Bruno, and Steve Woolgar. *Laboratory Life : The Social Construction of Scientific Facts*. Beverly Hills: Sage Publications, 1979.

Lee, Desmond, ed. *Wittgenstein's Lectures, Cambridge, 1930–1932: From the Notes of John King and Desmond Lee*. Oxford: Blackwell, 1980.

Lee, T. D., and C. N. Yang. "Question of Parity Conservation in Weak Interactions." *Physical Review* 104 (1956): 254–58.

Lévi-Strauss, Claude. *Structural Anthropology*. New York: Basic Books, 1963.

Little, Daniel. *A New Social Ontology of Government* Cham: Palgrave, 2020.

Livingston, Stanley M. *Particle Accelerators: A Brief History*. Cambridge, MA: Harvard University Press, 1969.

Locke, John. *An Essay Concerning Human Understanding*. Cleveland: Meridian, 1964.

Manetti, Giovanni. *Theories of the Sign in Classical Antiquity*. Bloomington: Indiana University Press, 1993.

Mangano, Michelangelo. "Understanding the Standard Model, as a Bridge to the Discovery of New Phenomena at the LHC." *arXiv:0802.0026v2 [hep-ph]* (2008): 15.

——. "Visible and Invisible in Modern Physics." CERN, Theoretical Physics Department, Arts and Science Across Italy, September 9–13, 2018.

Marcuse, Herbert. *One Dimensional Man: Studies in the Ideology of Advanced Industrial Society*. Boston: Beacon, 1964.

Marder, Jenny. "At Cern, Art Collides with Science." PBS. http://www.pbs .org/newshour/updates/science/july-dec10/cern_12-01.html.

Marx, Karl, and Friedrich Engels. *Capital: A Critique of Political Economy*. New York: International, 1967.

——. *Economic and Philosophic Manuscripts of 1844*, Great Books in Philosophy. Amherst, NY: Prometheus, 1988.

——. *The German Ideology: Including Theses on Feuerbach and Introduction to the Critique of Political Economy*, Great Books in Philosophy. Amherst, NY: Prometheus, 1998.

Marx, Karl, Friedrich Engels, Roy Pascal, W. Lough, and C. P. Magill. *The German Ideology, Parts I and III*. New York: International, 1947.

Masco, Jospeh. "Lie Detectors: On Secrets and Hypersecurity in Los Alamos." *Public Culture* 14, no. 3 (2002): 441–67.

——. *The Nuclear Borderlands: The Manhattan Project in Post–Cold War New Mexico*. Princeton, NJ: Princeton University Press, 2006.

Merton, Robert King. *The Sociology of Science: Theoretical and Empirical Investigations*. Chicago: University of Chicago Press, 1973.

Miller, Arthur I. *Insights of Genius: Imagery and Creativity in Science and Art*. Cambridge, MA: MIT Press, 2000.

Mlodinow, Leonard. *Feynman's Rainbow: A Search for Beauty in Physics and in Life*. New York: Warner, 2003.

Natoli, Joseph, and Linda Hutcheon, eds. *A Postmodern Reader*. Albany: State University of New York Press, 1993.

Nöth, Winfried. *Handbook of Semiotics*. Bloomington: Indiana University Press, 1990.

Needham, Rodney. *Right and Left; Essays on Dual Symbolic Classification*. Chicago: University of Chicago Press, 1973.

Noth, Winfried. *Handbook of Semiotics*. Bloomington: Indiana University Press, 1990.

Ortner, Sherry B. "Theory in Anthropology Since the Sixties." *Comparative Studies in Society and History* 26, no. 1 (1984): 126–66.

Paracelsus, and Arthur Edward Waite. *The Hermetic and Alchemical Writings of Aureolus Philippus Theophrastus Bombast, of Hohenheim, Called Paracelsus the Great*. New Hyde Park, NY: University Books, 1967.

Peirce, Charles S., Max Harold Fisch, Christian J. W. Kloesel, and Project Peirce Edition. *Writings of Charles S. Peirce: A Chronological Edition*. Bloomington: Indiana University Press, 1982.

Persico, Enrico, Ezzio Ferrari, and Sergio E. Segre. *Principles of Particle Accelerators*. New York: W. A. Benjamin, 1968.

Pestre, Dominique. "Commemorative Practices at CERN: Between Physicists' Memories and Historians' Narratives." *Osiris* 14 (1999): 203–16.

Pickering, Andrew. *Constructing Quarks: A Sociological History of Particle Physics*. Chicago: University of Chicago Press, 1984.

——. "The Hunting of the Quark." *Isis* 72, no. 2 (1981): 216–36.

——. *The Mangle of Practice: Time, Agency, and Science*. Chicago: University of Chicago Press, 1995.

——. "The Mangle of Practice: Agency and Emergence in the Sociology of Science." *American Journal of Sociology* 99, no. 3 (1993): 559–89.

Pinch, Trevor J. "Opening Black Boxes: Science, Technology and Society." *Social Studies of Science* 22, no. 3 (1992): 487–510.

Politzer, H. David. "The Dilemma of Attribution." In *Nobel Lecture*, December 8, 2004.

Rabinow, Paul. *Anthropos Today: Reflections on Modern Equipment*. Princeton, NJ: Princeton University Press, 2003.

——. *Essays on the Anthropology of Reason*. Princeton, NJ: Princeton University Press, 1996.

——. *Making PCR: A Story of Biotechnology*. Chicago: University of Chicago Press, 1996.

——. *Marking Time: On the Anthropology of the Contemporary*. Princeton, NJ: Princeton University Press, 2008.

——. "Prosperity, Amelioration, Flourishing: From a Logic of Practical Judgment to Reconstruction." *Law and Literature* 21, no. 3 (2009): 301–20.

Rabinow, Paul, and Gaymon Bennett. *Designing Human Practices: An Experiment with Human Practices*. Chicago: University of Chicago Press, 2012.

Ragain, Melissa. "From Organization to Network: MIT's Center for Advanced Visual Studies." *X-tra* 3 (Spring 2012). http://x-traonline.org /article/from-organization-to-network-mits-center-for-advanced-visual -studies.

Ranciere, Jacques. *Aesthetics and Its Discontents*. Cambridge: Polity, 2009.

Reichenbach, Hans. *The Rise of Scientific Philosophy*. Berkeley: University of California Press, 1951.

Rheinberger, Hans-Jorg. *Toward a History of Epistemic Things: Synthesizing Proteins in the Test Tube*. Stanford, CA: Stanford University Press, 1997.

Roy, Arpita. "Science and the Large Hadron Collider: A Probe Into Instrumentation, Periodization and Classification." *Dialectical Anthropology* 36, no 3–4 (2012): 291–316. http://dx.doi.org/10.1007/s10624-012-9278-6

——. "Ethnography and Theory of the Signature in Physics." *Cultural Anthropology* 29, no. 3 (2014): 479–502. http://dx.doi.org/10.14506/ca29.3

——. "Anthropology as an Experimental Mode of Inquiry." In *Anthropology and Ethnography Are Not Equivalent*, ed. Irfan Ahmad. New York: Berghahn, 2021.

Ruskin, John. *Modern Painters*. London, J. M. Dent, 1906.

Russell, Bertrand. *The Collected Papers of Bertrand Russell*. Ed. Kenneth Blackwell. London: Allen & Unwin, 1983.

Russell, Bertrand. *An Inquiry Into Meaning and Truth*. London: Routledge, 1992.

——. *Our Knowledge of the External World as a Field for Scientific Method in Philosophy*. London: Allen & Unwin, 1926.

Sample, Ian. "Is the Large Hadron Collider Worth Its Massive Price Tag?" *The Guardian*, September 22, 2009. https://www.theguardian.com/science/blog/2009/sep/22/particlephysics-cern.

Santayana, George. *The Life of Reason; or, the Phases of Human Progress*. New York: Scribner's, 1905.

Schatzki, Theodore R., Karin Knorr-Cetina, and Eike von Savigny, eds. *The Practice Turn in Contemporary Theory*. London: Routledge, 2001.

Sebeok, Thomas A. *Contributions to the Doctrine of Signs*. Bloomington: Indiana University Press, 1976.

Sessler, A. M., and E. J. N. Wilson. *Engines of Discovery: A Century of Particle Accelerators*. Hackensack, NJ: World Scientific, 2007.

Shapin, Steven. *The Scientific Life: A Moral History of a Late Modern Vocation*. Chicago: University of Chicago Press, 2008.

Shapin, Steven, Simon Schaffer, and Thomas Hobbes. *Leviathan and the Air-Pump: Hobbes, Boyle, and the Experimental Life: Including a Translation of Thomas Hobbes, Dialogus Physicus De Natura Aeris by Simon Schaffer*. Princeton, NJ: Princeton University Press, 1985.

Shlain, Leonard. *Art and Physics: Parallel Visions in Space, Time, and Light*, New York: Morrow, 1991.

Smidt, Posted by Joseph. "Http://Www.Theeternaluniverse.Com/2010/12/What-Would-Happen-to-Particle-Physics.Html." December 29, 2010.

Smith, C.H. Llewellyn. "Http://Public.Web.Cern.Ch/Public/En/About/Basicscience3-En.Html."

Sokal, Alan. "A Physicist Experiments with Cultural Studies." *Lingua Franca* (May/June 1996): 62–64.

——. *Beyond the Hoax: Science, Philosophy, and Culture.* Oxford: Oxford University Press, 2008.

Sokal, Alan, and Jean Bricmont. *Fashionable Nonsense: Postmodern Intellectuals' Abuse of Science.* New York: Picador, 1998.

Solovey, Mark. "Project Camelot and the 1960s Epistemological Revolution: Rethinking the Politics-Patronage-Social Science Nexus." *Social Studies of Science* 31, no. 2 (2001): 171–206.

Stewart, Ian. *Why Beauty Is Truth: A History of Symmetry.* New York: Basic Books, 2007.

Strathern, Marilyn. *Audit Cultures: Anthropological Studies in Accountability, Ethics, and the Academy.* London: Routledge, 2000.

——. *Partial Connections.* Savage, MD: Rowman & Littlefield, 1991.

Sutton, Christine. "Creativity Across Cultures." *CERN Courier* 58, no. 6 (July/August 2018).

Tarasov, L. V. *This Amazingly Symmetrical World.* Moscow: Mir, 1986.

Taubes, Gary. *Nobel Dreams: Power, Deceit, and the Ultimate Experiment.* New York: Random House, 1986.

Taylor, Charles. *Philosophy and the Human Sciences.* Cambridge: Cambridge University Press, 1985.

Todorov, Tzvetan. *Theories of the Symbol.* Ithaca, NY: Cornell University Press, 1982.

Traweek, Sharon. *Beamtimes and Lifetimes: The World of High-Energy Physicists.* Cambridge, MA: Harvard University Press, 1988.

Turner, Victor. "Social Dramas and Stories About Them." *Critical Inquiry* 7, no. 1 (1980): 141–68.

Uberoi, J. P. Singh. *The European Modernity: Science, Truth, and Method.* New Delhi: Oxford University Press, 2002.

——. *The Other Mind of Europe: Goethe as a Scientist.* Delhi: Oxford University Press, 1984.

——. *Science and Culture.* Delhi: Oxford University Press, 1978.

Walsh, Lorraine. "Art and Outreach." *SCGP News,* Special Issue XV (2020): 70–72.

Weber, Max. "Science as a Vocation." *Daedalus* 87, no. 1 (1958): 111–34.

Weeks, Andrew. *Paracelsus: Speculative Theory and the Crisis of the Early Reformation.* Albany: State University of New York Press, 1997.

Weinberg, Steven. *Dreams of a Final Theory.* New York: Pantheon, 1992.

——. *Facing Up: Science and Its Cultural Adversaries.* Cambridge, MA: Harvard University Press, 2001.

——. "Viewpoints on String Theory: Steven Weinberg." NOVA, July 2003. http://www.pbs.org/wgbh/nova/elegant/view-weinberg.html.

Weyl, Hermann. "Gravitation and the Electron." *Proceedings of the National Academy of Sciences of the United States of America* 15, no. 4 (1929): 323–34.

——. *Mind and Nature Selected Writings on Philosophy, Mathematics, and Physics.* Princeton, NJ: Princeton University Press, 2009.

——. *Symmetry.* Princeton, NJ: Princeton University Press, 1952.

Whitehead, Alfred North. *Science and the Modern World. Lowell Lectures, 1925.* New York: Macmillan, 1925.

Wigner, Eugene P. *Symmetries and Reflections: Scientific Essays of Eugene P. Wigner.* Bloomington: Indiana University Press, 1967.

——. "The Unreasonable Effectiveness of Mathematics in the Natural Sciences." *Communications in Pure and Applied Mathematics* 13, no. 1 (1960): 1–14.

Wilczek, Frank. *The Lightness of Being: Mass, Ether, and the Unification of Forces.* New York, NY: Basic Books, 2008.

Wilczek, Frank, and Betsy Devine. *Longing for the Harmonies: Themes and Variations from Modern Physics.* New York: Norton, 1987.

Wittgenstein, Ludwig. *Culture and Value.* Trans. Peter Winch. 1961. Reprint, Chicago: University of Chicago Press, 1984.

——. *Philosophical Investigations.* New York: Macmillan, 1953.

——. *Tractatus Logico-Philosophicus.* Trans. David Pears and Brian McGuinness. London: Routledge, 2001.

Wittgenstein, Ludwig, G. E. M. Anscombe, Rush Rhees, and G. H. von Wright. *Remarks on the Foundations of Mathematics.* Cambridge, MA: MIT Press, 1967.

Woit, Peter. *Not Even Wrong: The Failure of String Theory and the Search for Unity in Physical Law.* New York: Basic Books, 2006.

Wroblewski, A. K. "The Downfall of Parity: The Revolution That Happened Fifty Years Ago." *Acta Physica Polonica* 39 no. 2 (2008): 251–64.

Wyatt, Terry. "High-Energy Colliders and the Rise of the Standard Model." *Nature* 448, no. 7151 (2007): 274–80.

Yates, Frances Amelia. *Giordano Bruno and the Hermetic Tradition.* Chicago: University of Chicago Press, 1964.

INDEX

GPSR Authorized Representative: Easy Access System Europe, Mustamäe tee 50, 10621 Tallinn, Estonia, gpsr.requests@easproject.com

www.ingramcontent.com/pod-product-compliance
Lightning Source LLC
Chambersburg PA
CBHW021853020426
42334CB00013B/315